Studying

Palgrave Study Skills

Business Degree Success
Career Skills
Cite Them Right (8th edn)
Critical Thinking Skills (2nd edn)
e-Learning Skills (2nd edn)
The Exam Skills Handbook (2nd edn)
Great Ways to Learn Anatomy and Physiology
How to Begin Studying English Literature (3rd edn)
How to Manage Your Distance and Open Learning
  Course
How to Manage Your Postgraduate Course
How to Study Foreign Languages
How to Study Linguistics (2nd edn)
How to Use Reading in Your Essays
How to Write Better Essays (2nd edn)
How to Write Your Undergraduate Dissertation
Information Skills
The International Student Handbook
IT Skills for Successful Study
The Mature Student's Guide to Writing (3rd edn)
The Mature Student's Handbook
The Palgrave Student Planner
Practical Criticism
Presentation Skills for Students (2nd edn)
The Principles of Writing in Psychology

Professional writing (2nd edn)
Researching Online
Skills for Success (2nd edn)
The Student's Guide to Writing (3rd edn)
Study Skills Connected
Study Skills for International Postgraduates
The Study Skills Handbook (3rd edn)
Study Skills for Speakers of English as a Second
  Language
Studying History (3rd edn)
Studying Law (3rd edn)
Studying Modern Drama (2nd edn)
Studying Psychology (2nd edn)
Teaching Study Skills and Supporting Learning
The Undergraduate Research Handbook
The Work-Based Learning Student Handbook
  Work Placement – A Survival Guide for
  Students to Work Placements – A Survival
  Guide for Students
Write it Right (2nd edn)
Writing for Engineers (3rd edn)
Writing for Law
Writing for Nursing and Midwifery Students
  (2nd edn)
You2Uni

Pocket Study Skills

14 Days to Exam Success
Blogs, Wikis, Podcasts and More
Brilliant Writing Tips for Students
Completing Your PhD
Doing Research
Getting Critical
Planning Your Essay
Planning Your PhD
Reading and Making Notes

Referencing and Understanding Plagiarism
Reflective Writing
Report Writing
Science Study Skills
Studying with Dyslexia
Success in Groupwork
Time Management
Writing for University

Palgrave Research Skills

Authoring a PhD
The Foundations of Research (2nd edn)
The Good Supervisor (2nd edn)
The Postgraduate Research Handbook (2nd edn)
Structuring Your Research Thesis

For a complete listing of all our titles in this area please visit **www.palgrave.com/studyskills**

Studying Physics

David Sands

First published 2004 by
PALGRAVE MACMILLAN
Houndmills, Basingstoke, Hampshire RG21 6XS and
175 Fifth Avenue, New York, N.Y. 10010
Companies and representatives throughout the world

PALGRAVE MACMILLAN is the global academic imprint of the Palgrave
Macmillan division of St. Martin's Press, LLC and of Palgrave Macmillan Ltd.
Macmillan® is a registered trademark in the United States, United Kingdom
and other countries. Palgrave is a registered trademark in the European
Union and other countries

ISBN 1–4039–0328–X

This book is printed on paper suitable for recycling and made from fully
managed and sustained forest sources.

A catalogue record for this book is available from the British Library.

10   9   8   7   6   5   4   3   2   1
13   12   11   10   09   08   07   06   05   04

Printed and bound in UK

To Tomomi, Taro and Chibi

Contents

Some of the common problems encountered by undergraduate physicists are described, together with their implications for learning physics. The concept of physics as a problem-solving, or research, process is developed, leading to identifiable critical thinking skills needed in experimental and theoretical physics

The basic mathematical techniques needed to analyse experiments are developed, including common mathematical functions, graph plotting, curve fitting, and elementary calculus.

The physics of a particular problem influences the experimental method adopted and this chapter discusses the options open to the experimenter, especially in respect of the nature and sources of experimental errors. The principles of electronic instrumentation are also described.

Probability is an important part of physics, and the most common probability distributions used by physicists are developed in detail. In particular, their use in the analysis of experimental data and experimental errors are described.

List of Tables and Figures

Tables

Figures

Preface

The transition from A-level to degree level is difficult enough in any subject, but in physics it is especially so. Perhaps it is becoming more difficult. The material in this book is designed to help you to make that transition by improving your learning and acquiring a range of skills required by the physicist but which students, for one reason or another, increasingly have difficulty with. In a recent discussion I had with one of our graduates who is now teaching in Hull, for example, it emerged that 60 per cent of A-level physics students in his school are not studying maths. If these students go on to university to study physics they will need to learn maths that would once have been learnt alongside their A-level physics. Such students are starting from a different position compared with students just a few years ago, so there is a good chance that not only will their mathematical abilities at the end of the course be reduced in comparison, but the shorter time in which they have to develop their understanding may well impact upon other aspects of their studies.

There is thus a need for a book about skills in physics. Unlike most textbooks, this book does not contain different chapters devoted to specific topics such as mechanics, optics, heat and so on. Instead it contains chapters on the mathematics of experimental physics, the design of experiments, statistics in physics, the role of theory, mathematical modelling, and scientific reporting. In short, the most essential skills, the 'how' and 'why' of physics, are described. These are the most difficult skills to acquire because they are normally hidden within a course and developed through experience; experience in practical physics gained in the laboratory, experience in theoretical physics gained in lectures, and experience in mathematical physics gained through solving problems. However, you can't learn these skills unless you have first identified them. This book is therefore intended to help you to identify these essential skills in physics at the outset, so that you can spend your valuable time concentrating on what is important and hence make the most of your undergraduate experience.

This material is supplementary to your normal classes in lectures and laboratories and ideally it should be read within your first year. Some chapters will be easier to understand than others and you may need to revisit parts, depending on your experience. You may need to write a report, for example, and have it marked critically, before you can compare your approach with that described here. Having made a mistake it is often easier to see where you went wrong than it is to try to absorb the principles at the outset and avoid the mistakes altogether. If you read any part of the book and find it difficult to understand or to relate to your own personal experiences then simply leave those parts until a later stage in your studies. There is no harm, and probably much benefit, in re-reading the book in your second, or even your third, year. Even if you understand the book first time around your views will change with experience and so will your perceptions and interpretation of many of these ideas. It will be stressed throughout this book that learning is an activity; it is also a life-long activity. One of the marvels of physics is that no matter how long you have been working there is always something new to be learned, even about things you might consider yourself knowledgeable.

DAVID SANDS

Acknowledgements

I would like to thank my colleagues in the physics department at Hull for their support. In particular, Drs Phil Key, Chris Hogg and Jeremy Dunning-Davies have offered valuable help and suggestions. I would also like to thank my colleague and friend, Dr K. M. Brunson, of Qinetiq, with whom I have worked over the years, and who has made valued comments on some of the contents of this book. Paul Monk's contribution is especially welcome.

DAVID SANDS

1 Studying Physics: An Introduction

Have you ever found yourself telling a complete stranger at a social gathering that you study physics? If you have I'll bet the reaction was not enthusiastic. Quite likely there was an awkward silence, perhaps even it signalled the end of that conversation. This is normal, I'm afraid. I knew someone once who avoided this difficulty by describing himself as a quantum mechanic, but I can't imagine that it really solved the problem. It might surprise you to learn, then, that physicists are highly sought after in the job market. A PhD in physics can earn a fortune in the financial institutions in the City, for example. Why? Because of the skills possessed by the physicist. You might think that financial institutions would want people trained in financial matters, who can analyse market trends and predict likely profitable future markets. Physicists have no such training but could tell you all you need to know about nuclear power or quantum mechanics or planetary motion. What could they bring to such a job? You would be surprised. Physicists are not only skilled in mathematics, but also in applying mathematics to physical situations. Analysing markets is just another application of mathematics, though one subject to considerably less certainty than the situations normally dealt with by physicists.

It is the intention of this book to describe in detail the skills possessed by the physicist. Among other things, you will need to be able to:

- Understand mathematical arguments
- Relate mathematics to physics
- Design experiments or formulate theory
- Analyse experimental data
- Understand significance and errors
- Perform statistical analysis
- Evaluate the impact of new data on existing theories.

These skills encompass a range of mathematical and analytical skills that can also be applied to a range of activities outside physics. Employment in economic and financial fields is but one option.

▶ Learning skills

The idea of identifying skills and presenting them in such a manner that a physics undergraduate can learn and apply them is relatively modern. There was a time, not too far back in the past, when an undergraduate simply studied physics and picked up the skills along the way. Such was the case when I went to university in 1978. I well remember in my second year of study struggling with the art of writing a laboratory report; sometimes I would get good marks, at other times bad, but always with the same effort from me. Clearly whatever I was doing to get the good marks was not systematic, and I needed to know why. I sought advice from a senior academic in the department who suggested I should read a scientific paper and try to pick up some hints about style. I duly went to the library only to find myself confronted by row upon row of periodicals. Where to start? I chose one at random and opened it at random. I found the physics within so far beyond me that I gave up that idea. I learnt most about writing reports in my third year during an industrial placement from a very helpful man who sat down with me and went through what I had written page by page. To him I owe a great deal.

The undergraduate education system clearly had its flaws, but it worked to an extent nonetheless. As with countless others over the years I duly graduated and moved into a career. I developed my report-writing skills as well as many of the other skills I shall write about in this book without, in many cases, being particularly aware of them. Such a system would not work today, however. University education in Britain in general has changed substantially, driven in part by the increasing numbers entering the system and in part by the changes in school education. It is not that students are less able, but there is a wider range of abilities in education as a result of the expansions that have taken place and the entry level skills of students have also changed. There is even talk in some places of the 'maths-less' physics degree.

To a great extent these are problems for the educators; how to design a curriculum to deliver particular knowledge and skills in the face of these changes and challenges. However, it is also a problem for the student. You will find that upon entering a course of study you will be faced with more choice over your subjects of study than ever I was allowed, and that in many institutions it is possible even to choose a number of non-physics elements or modules. Of course there will always be a compulsory element of core material. It would be a very badly designed course indeed that would allow you to opt out of such subjects, but it is one thing to attend lectures, quite another to learn and develop a range of skills. It has always been a sub-text to higher education that students study to pass exams rather than to learn. After all it is the final degree result that matters most. It is better to get a first-class honours degree rather than a second, a second rather than a third, and a third rather than a pass. It is only natural that as a student you will pay attention to those things that will get you the highest marks, including choosing 'soft' options among your modules. Taken to the extreme, you may well end up with a good degree and a good job but find yourself at some point wishing you had paid more attention to such and such. Too late! The opportunity has passed.

A good degree result is important, no denying that. A first or upper-second will open doors that a lower-second will not, but within a short time of graduating the importance of the degree result will begin to fade against your achievements since then. In short, your ability to do your job will be the criterion against which you are judged, and this will depend not on your degree classification *per se*, though of course it should ideally reflect your abilities, but on the manner in which you apply yourself to the task in hand using the skills you have acquired. As a student you owe it to yourself to maximise your return from your education. You can do this by taking responsibility for your own learning so that not only do you emerge from your studies with a good degree but with an armoury of skills at your disposal.

The skill of learning

Taking responsibility for your own learning does not mean that you should ignore the curriculum or, worse, that the curriculum you are presented with at your particular institution of learning is not worth bothering with. Most physics departments in the UK have been judged excellent by the Quality Assurance Agency and most will deliver a good course. That doesn't mean to say, though, that you will necessarily learn much. Learning is an activity; you, the learner, must do something to learn. The following story from one of Richard Feynman's autobiographies illustrates the point.

Richard Feynman was one of the later twentieth century's most cele-brated physicists. After graduating, however, he temporarily had a poor opinion of his own abilities after confiding in his sister during one visit that he was unable to understand the work he was supposed to be doing. It was just after he had taken up his (post)graduate work and he was read-ing a paper on his particular field. He told his sister that those who had written it must have been geniuses because he knew that he was able (his first degree results told him that) but he couldn't understand a word of it.

His sister replied, seemingly without much sympathy, that it was because he wasn't studying it. Of course Richard Feynman rejected this idea. He had been studying for years. He knew about physics, but he couldn't get to grips with this paper. His sister was adamant. He wasn't studying it as he would have done an undergraduate text. One thing about Richard Feynman was that he was a humble man. He accepted what his sister had to say and approached the paper in a different way, taking it apart piece by piece, equation by equation. To his surprise he found that he started to make progress and very soon understood what previously had appeared impenetrable.

Richard Feynman had forgotten that learning is an active process. Blinded by his own achievements up to that point he believed that he no longer needed to study, and that simply by reading he would understand. It took his sister, who must have been a very wise woman to see his mistake so easily, to point out to him that he needed to return to the basics, to concentrate on those things which had allowed him to get to his present position. That is to say, he needed to study to understand.

All of us make this mistake at some time. Our reasons may be different from Richard Feynman's but the result is the same. Instead of understanding and learning we acquire information and think we understand, only to find at the crucial time when we need to apply the knowledge that we cannot. School is supposed to prepare us for this. Increasingly it appears it does not. In the modern age we are overloaded with information. It is on the radio, on television, in the newspapers and magazines, and on the internet. Given an assignment to write about nuclear power, say, we could find enough information to fill several books. We need to distil it. We need to extract from all that we have acquired the essential message we wish to deliver and put it across coherently. No problem, you might say; it's something of which you have considerable experience. Assignments and coursework are an increasing part of the school curriculum. How much of it, though, have you really understood? How much of it have you *studied*?

The answer is probably very little, because you don't need to. It is relatively easy to collect and collate information and present it well without going beyond a superficial understanding. How many times in such an assignment would you stop to really consider a point that you don't understand? Perhaps you would opt simply to leave it out but you might include it anyway on the assumption that you won't be questioned in detail about it. Not only do we fail to learn by doing this – in some instances consciously so – but if we were to treat other learning situations in the same manner we would fail to learn on those occasions too.

This, I believe, is the biggest challenge facing the modern student. It seems that more and more students do not know how to learn, and in a subject like physics, which takes a substantial intellectual effort to grasp with each level building securely on the foundations of the previous, a failure at the earliest stages will feed through to the highest levels of your study. Learning is an activity, and you the learner have to undertake it. It doesn't matter whether your lecturers are good or bad, whether they excite your interest or whether they are deadly boring, whether they use the blackboard or multimedia powerpoint presentations; once the information is delivered to you, you have to deal with it. If it is not structured in a way that is easy to follow you have to restructure it. If there are contradictions you have to resolve them. You have to take responsibility for your own learning.

▶ Undergraduate physics: a unity of ideas

Undergraduate physics degree programmes vary widely in their content and structure. Traditionally the higher levels of the degree programme, such as the third or fourth year, will contain large amounts of material that reflect the research interests of the departmental staff. At lower levels there will be a great deal of material that will be common to physics courses around the country. Thus you will find classical and quantum physics comprising, among other things:

- mechanics
- electricity and magnetism
- solid-state physics
- optics
- thermodynamics
- laboratory classes
- quantum mechanics
- relativity

These seem to be distinct and separate subjects but in fact they share many ideas in common. Quite when a student begins to appreciate this point is unclear, but as demonstrated in Chapter 5, it is an important point to appreciate. The way that you learn will change from accumulating facts and figures appropriate to separate subjects by comparing, contrasting, assimilating and synthesising, and your ability to understand new physics concepts may well also change.

▶ Undergraduate mathematics: the glue that binds

Mathematics is usually taught separately in conventional physics degrees, but maths is the language through which we express our ideas of the natural physical world. Mathematics is a common thread that runs through the entirety of physics, and seemingly when there is no direct connection between different concepts in physics there may well be a mathematical connection. This topic is explored in greater detail in Chapter 6.

Undergraduate physics requires fairly detailed maths, and the courses will usually cover at least:

- algebra
- complex numbers
- vectors
- differentiation
- integration
- differential equations
- matrices

You may well find more advanced topics than these, such as Fourier series, Laplace transforms, Legendre polynomials and, among the physics components, statistical mechanics and nuclear and particle physics, to name but two.

As it stands these two lists describe an average physics department, but it is not comprehensive. Some of these topics are described in this book, but there isn't space to cover everything. Likewise with a physics degree. Physics is such a vast subject that it is impossible to teach it all and among departments there will be variations in what is taught and how. As an average, however, it also provides a bald statement of the academic experience of the average student, if there is such a thing, and you can expect to find something very similar.

▶ The undergraduate experience: identifying the problems

Of course, the undergraduate experience is made up of much more than the list of subjects given above. There is the whole field of your social development as well as academic, but my concern lies solely with the latter. Much more important than the list of subjects is how you cope with them. Let me summarise a typical student's experiences in physics. Many students often do find the transition from sixth form to first year at university difficult. The amount of new material can be overwhelming. Students do not always understand what they are expected to do with the lecture material given to them and sometimes look enviously across at other disciplines where the material is of a different nature and, apart from there being less of it, perhaps easier to understand. Students especially have difficulty writing laboratory reports, and increasingly are having difficulty with the mathematics content of physics courses. Come exam time, the tendency to question-spot and to memorise often leads to a patchy, and occasionally disastrous, performance.

Experience and expectations

There are a number of issues I wish to identify as especially important. First among them is the question of expectations; what is it you expected would face you at university? Choosing an appropriate undergraduate degree programme has always contained a strong element of luck. Surely not in physics, you might think; physics is such a difficult subject that only somebody who was confident of their interest would study it, and such people would clearly be realistic in their expectations? Not so. You might find physics at school interesting but have no comprehension that studying it to degree level would be worthwhile. There is no physics industry as such, and physics degrees are not associated with particular professions in the way that degrees in law or psychology, or business and management, to name a few, are. Some will study it because they have a strong interest in it, but there will be those who drift into physics because they are, or were, reasonably able at school and studying physics is as good as studying anything else. You may have certain expectations about the subject based on your experiences at school. Should you find that reality does not match your expectations you may well struggle. One way of overcoming such a struggle is to be clear about what it is that is required of you and what it is that you are trying to achieve.

You may not be aware that the problems you face are caused by your own expectations. For example, you may be convinced that your problems are not of your making; that they arise, for example, from ineffective lecturing or a lack of concern among staff. You could change departments, but that's a drastic step. Much more practical and worthwhile is to overcome the effects of 'bad' teaching through your own efforts. As I have indicated, bad teaching is really quite hard to define. Some lecturers are more interesting than others and some will structure their material better than others, but what this means in reality is that the amount of effort you have to put in to structure your notes in such a way that they are comprehensible to you now, and useful to you for the remainder of your course, will vary from subject to subject, and from lecturer to lecturer. One thing is clear, though. You will always have to work. Very, very few people can absorb knowledge without seeming to make an effort. The rest of us have to undertake the activity of learning and to acquire the skill of learning.

Making the most of the teaching

As I have already indicated, different departments will go about delivering their courses in different ways. There will be a considerable commonality, however, across departments because the nature of the subject influences the teaching. Most teaching will be delivered by conventional lecture, but even among lecturers there are variations in style. In a 2002 article entitled, 'How do we Know if we are Doing a Good Job in Physics Teaching?' Robert Ehrlich, of George Mason University, Virginia,[1] talks enthusiastically about the innovations he has made in teaching only to report that some students like it and some don't. And this is the essential difficulty. Methods of learning and studying vary among individuals and whilst it is possible to identify methods and practices that suit some, there is no guarantee that they will suit others. Learning is an individual activity. The methods you employ to study and to learn must suit you.

There will be a lot of new material, even in the first year, and how you cope with this information is crucial to your learning. Is it enough, for example, to be able to make good notes during the lecture and to be able to learn them when revising? What is meant by learning in this context? What do you do if you are uncertain what to revise? What sort of additional material will be provided to support the lectures? How is it that other departments can, and do, adopt different approaches to learning, by for example holding seminars and tutorials to discuss what has been gained in the lecture, when physics departments invariably do not? There will be tutorials, but these will often not be the same as in other departments where a subject matter is specified, a particular article or chapter is read prior to the tutorial, and at the tutorial itself the topic is discussed. In fact you will find very little discussion of physics outside lectures; most of the time the subject will be presented as 'fact'. You have to recognise the nature of the subject and devise the most appropriate learning strategy for you.

▶ Physics as fact

The factual nature of much of physics teaching is sometimes difficult for students to understand. Again it is a question of expectations. Many students entering university will have encountered the idea that there is no such thing as a 'right' theory; there are theories that work and those that don't, but of those that work we cannot call them 'right' because who knows what developments in future may cause us to change our minds. This is none other than an expression of the philosopher Karl Popper's views on falsifiability in science; we can never prove a theory right, we can only prove it false. It's quite a popular view, not least because it sounds reasonable and describes to a great extent the historical development of many branches of science, not just physics. The problem for the student lies in relating this view to what is encountered in the lectures.

Physics has been taught in universities for many years and the methods used have largely stood the test of time. These methods are characterised by a factual approach to the subject that does not lend itself easily to discussion. There are foundations in physics that must be mastered before you are able to contribute meaningfully to the subject, and these foundations do not change appreciably with time. Much of thermodynamics was developed in the last century and the material taught to undergraduates will not, in all probability, have changed much in that time, except perhaps to decrease in importance as new material, such as solid-state physics, has entered the curriculum. Even basic solid-state physics has not changed much since the 1930s and 1940s; it is principally the technology, for example the development of modern semiconductor devices and circuits, that changes rapidly, and as an undergraduate you might not expect to deal with this in any depth until your final year. Even where fundamental physics changes, as in particle physics, much of what is taught will be presented as fact.

There will be very little, or no, discussion on the nature of the knowledge and scientific knowledge in particular. As an undergraduate you can expect to meet people with a very wide range of interests, including those engaged in philosophy. Such people may believe that all knowledge is subjective and the certainty that characterises physics is false. You may already have met these ideas. If and when you do, it should not alter your perceptions of physics or your approach to your studies. Physics is one of those few fields of human endeavour where the term 'classical' is used to express outmoded ideas. In music or literature, for example, classical implies a timeless quality associated with a basic goodness or worth. Not in physics! but that doesn't mean to say the ideas no longer have worth. Quantum mechanics may have replaced classical mechanics as the essential paradigm, but those areas of activity that were once exclusively and effectively described by Newtonian mechanics are probably still better described by it than by its successor. Newtonian mechanics is not 'wrong' in this sense. It is still very relevant and has about it a certainty and concreteness. The entire industrial development of the world, at least prior to the advent of electronics, was based on the principles of Newtonian mechanics in conjunction with classical thermodynamics. The reality of the machines or the age they ushered in cannot be denied nor can the applicability of the classical approach. The fact that it does not apply to other circumstances is another matter.

The principles of classical mechanics are true in an objectively quantifiable way and you would rightly expect this notion of factuality to inform the teaching of the subject, so that basic physics principles are taught very much as facts about which there can be no argument. This is not a deficiency in the teaching method but a direct consequence of the nature of the subject. The question of your having your own opinions, as you might be expected to in, say, history, economics or law, is irrelevant. It may be that you do not understand what you have been taught, but that also is another matter. Not surprisingly, then, you will not find much philosophical discourse on the nature of physics outside the lecture theatres. It has to be recognised, though, that there are areas of physics where the theories are not complete and where you might be able to form opinions of your own, but you are very unlikely to meet those until the later years of the course.

▶ Skills in physics

There are a number of issues I have mentioned which impact greatly on our quest to identify the skills required of a physicist. First, the laboratory classes: what is the purpose of laboratory classes; how does work in the laboratory, which initially is very directed, allow you to develop your skills to the point where you can undertake an open-ended research task such as a final-year project? Presentational skills are an important aspect of laboratory work. Whether scientific information is presented in written or oral form it must be presented lucidly and at a level commensurate with technical abilities of the intended audience. What constitutes a good laboratory report? Why, in my own case and for many others, was hard work alone not sufficient to ensure a good mark? Skill in presenting technical material is essential to the modern physicist and there are general rules for good practice which I shall describe in Chapter 7.

Other equally important skills include mathematics, IT and computational skills. Mathematics has already been mentioned and should be self-evidently important. Without it theories cannot be quantitatively predictive and we have no means by which to analyse experimental data. However, you might find that your mathematical skills do not match the requirements of a particular module or course element, and though the lecturer might be very helpful and willing you just cannot get to grips with the subject. Again, the question of your response to these situations is paramount. More than that, however, you should have a clear understanding of the role of maths in physics, a topic dealt with in Chapters 5 and 6. Computational skills include not only programming but also information technology in general. Spreadsheets, word-processing, graph plotting, scientific software, data acquisition and electronic sources of information, for example, are all important aspects of computer technology which aid the physicist. Added to which there are developments in computer aided learning (CAL) which you might wish to use, though I emphasise again, in CAL it is not the quality or quantity of information delivered to you that's important, rather how you use it.

Learning or simply passing exams?
Not least among the issues related to learning there is the question of assessment. It is natural for the assessment to skew your objectives in learning, but you should ask your self whether assessment is just about setting hurdles or is there some deeper purpose to it? You need to gain good marks, no question about it, but is there a conflict between your working towards assessments and developing skills? It is my contention that there is not. Granted, learning material in such a way that in can be recalled readily in the examination is not the same as learning something in a way that will be useful in future years. Very often we will find just a few weeks after sitting an exam that all the material we had crammed for it has been forgotten, seemingly beyond recall. It is useless beyond any purpose other than passing that exam. However, the ability to cram and recall is not sufficient in physics. You need to understand as well as memorise. Acquiring the skills to be an effective physicist is part of the process of understanding.

Most of the material from which your examinations will be drawn and from which you need to revise will come from lectures. It is imperative that you have a good set of notes, which is not as easy as it sounds. You may not notice it, but your concentration will lapse during a lecture; it will be virtually impossible for you to take in everything that is said, even if you can follow the arguments. The advice that I was given as an undergraduate consisted of the following: concentrate on writing during the lecture unless you are told to listen, but when writing don't worry about the neatness of your notes; just make sure they are legible, and within the next two or three days, while they are still reasonably fresh in your mind, transcribe them neatly to a notebook. I found it invaluable.

In this way not only do you remind yourself of the lecture content, but in writing out the notes you will be aware of any gaps where your concentration lapsed and your notes are not so easy to follow. Thus you can consult the textbooks and supplement your lecture notes with additional material, including mathematical steps that may have been missing in the lecture, or that you think you might need to recollect when you revise. In addition, you will have a reasonable grasp of the topic as you go so that the next lecture is easier to follow. You will find, if you haven't already, that if you begin to get lost during a lecture course subsequent lectures become harder to follow, and learning becomes very difficult.

This technique is far superior to attempting to make a decent set of notes during the lecture, a question posed earlier in this chapter. If you do this the attention you put into writing will not be given to the speaker, nor will you be aware of the deficiencies in your notes until you come to revise, when it is too late to do anything about it. Moreover, by not refreshing your memory shortly after the lecture you will not be prepared for the next instalment. Learning is thus seen to be a continuous activity rather than something that is done towards the end of the semester. There will always be an element of memory in successful exam technique. You have to be able to recall facts and formulae, and in the case of derivations, the steps that lead from A to B. There is no alternative to sitting down several days before an exam and cramming; memorising key concepts, ideas and facts. I can assure you, however, that if you understand the physics first, and have learnt it sufficiently well, you will do far better than if you concentrate on rote learning of formulae and hope that the right question comes up.

▶ Learning physics

Learning physics, then, involves a great many different skills. There is the skill of learning, of using your resources well, as well as the different skills in maths, experimental physics, computation, and presentation already mentioned. There are even some philosophical issues, such as falsifiability, the nature of theories, and the apparent factuality of physics. All these impinge in some manner on your learning, even philosophy. I say this because not many physicists regard the philosophy of science as having much useful to say in the practical pursuit of the subject, where the nature of a theory or the purpose of experimentation play second fiddle to the mastery of the technicalities and the detailed maths and physics of the problem. However, you are interested in more than just the detail. You are aiming to develop a range of skills that will enable you to be an effective physicist, and you need to appreciate how those skills fit together. I would suggest an awareness of the philosophical issues surrounding physics may help. This is dealt with in greater detail in Chapter 5 on the role of theory in physics, where the point is also made that a student's own perceptions of physics affects the approach to learning. The remainder of this chapter is therefore concerned with developing a view of physics in which the value of these skills is apparent and which will help you to develop an effective strategy for your own learning .

How is physics done?

It is necessary to identify some pattern, some process to physics, otherwise it becomes no more than a black art, its purpose and method to be discerned through experience and practice. If this is the case then this book has no purpose, for it becomes impossible to identify skills and describe them. This is clearly not the case, but it is the case that there is no universally agreed method. That is to say, not all physicists would agree on the ultimate aims, and hence the methods, of physics. I offer up the notion of physics as a problem-solving process. This is not so much a grand philosophy as a pragmatic approach to the description of the methods and the processes involved.

This definition will be a little too pragmatic for some who see in physics an elegance and simplicity in its description of the world that goes beyond mere problem-solving. It depends of course on what you mean by 'problem'. I mean it in the widest possible sense, so that even the subject of theoretical enquiry is seen as a problem. Physics is not just building things or solving technological problems, though that is of course very relevant to the industrial physicist or consultant. Nor is it engineering, though significantly virtually every branch of engineering has physics at its heart, including chemical engineering. Of course physics is concerned with describing and investigating the physical world, and may even be regarded as a branch of natural philosophy, but I suggest that for a practical view of its workings we regard it as a problem-solving process.

Physics as problem-solving

Restricting the discussion to the specific issue of undergraduate physics, but accepting that what is said is applicable far beyond that, it is possible to examine what conducting an experiment might be expected to achieve. Experimentation is the bedrock upon which physics is built. In the popular view theoreticians are more highly regarded, and certainly theoretical physics has kudos by virtue of the intellectual effort required, but a theory ultimately stands or falls on the experimental evidence. If there had been no experimental evidence gathered in support of Einstein's relativity theory, for example, then no matter how elegant it appears we would by now regard it with some scepticism. If we are to describe the natural physical world then it is to the world we must look. In short, if a theory does not match the experiment we are more likely to tear up the theory than to describe the experiment as inadequate, provided the experiment has been conducted properly.

We start, therefore, with experiments. Undergraduate laboratory experiments can be viewed as training exercises designed to equip you with a number of skills you will need as a professional physicist. Some of these skills will be technical; an appreciation of instrumentation, of errors, of significance, and so forth. Others will be concerned with the development of more general skills, loosely referred to as 'critical thinking skills'. The phrase is not mine, being very popular in a wide range of activities, especially business management. Critical thinking skills are emphasised very much as one of the essential tools in the manager's resource kit, and it is probably much more the case in a subject such as management that they need to be identified and taught separately. In physics these skills are an inherent part of the job so you should develop them naturally as you progress along your undergraduate physics career. Put very simply, if you don't have these skills you won't be a good physicist, but there is no harm in providing a helping hand by defining and describing the skills involved.

▶ The research process

The interest in describing these skills is primarily to show you how experimental and theoretical work can be seen formally as an exercise in solving a problem. Once you understand that you will have a better appreciation of what it is you are trying to achieve. Understanding the process will not necessarily help you with the mechanical skills of conducting the experimental work – these are dealt with in Chapters 2 and 3 – but having developed an awareness of the process you will be in a position to appreciate the important aspects to be included in a formal laboratory report or paper, as opposed to a daily laboratory diary. The research process can be represented as in Figure 1.1.

Critical thinking
The critical thinking skills required for the research process can be defined as:

1 *Gaining knowledge*: who has done what, where, and when? Knowledge is required to define the problem.

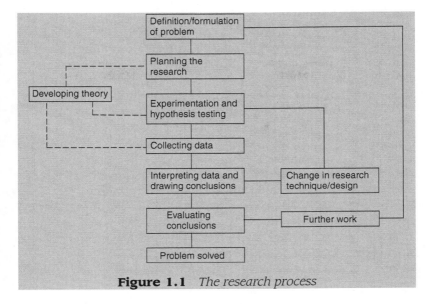

Figure 1.1 *The research process*

2 *Understanding*: describing past work, comparing different findings, contrasting different theories and explanations, explaining the nature of the problem to be solved. Understanding is essential to the formulation of the problem.

3 *Application of knowledge and understanding*: planning the research and developing the experimental method.

4 *Analysing*: interpreting data, asking what it means.

5 *Synthesizing*: drawing conclusions requires that separate items of information (results) be drawn together and new knowledge (conclusions) synthesized.

6 *Evaluating*: was the research successful? was the original problem solved? what else did you find? are more experiments needed?

▶ The scientific method

Written thus, with the critical thinking skills formally identified, the research process is seen not only to be logical, but also sensible. In fact it describes what is commonly known as the scientific method. Science today is taken to be a body of knowledge: thus we speak of the science of physics, or of chemistry, or of biology, but these are later developments. In its original meaning science describes a method. Sherlock Holmes is referred to as the first scientific detective, not for his specialist knowledge in particular specialities, but for his method of collecting data, of analysing and deducing. Specialist knowledge he had in abundance, but only in so far as it was useful to him; the plant biology of alkaloid poisons, the chemistry of perfumes, and so on. Anything that might allow him to analyse the evidence at a crime scene, in short, but not knowledge for knowledge's sake.

Science is a method. It is a logical, systematic method of investigation in which fancy has no place. Guesswork is allowed, for sometimes information is incomplete and an informed leap of the imagination is required. It has to be informed guesswork, though, and not just the product of an imagination run free. There are plenty of examples in the stories of Sherlock Holmes of an intuitive leap, but no matter; the stories adequately convey the notion of a detective who relies primarily on reason and logic backed up by a prodigious relevant knowledge. Sherlock Holmes may have been a detective but he was also a scientist, and his method is essentially the same as described in the research process above.

The scientific method in undergraduate physics

However, it is surprising how many undergraduates are unaware of exactly what it is they are doing when learning physics and what it is they are trying to achieve. An undergraduate degree programme is so structured as to provide not only the relevant background knowledge that you will need but also the skills required to perform research in a scientific manner. These skills should *normally* be acquired during the course of a degree programme, particularly as in the later years the experimentation becomes more like true research and less like well-defined experiments. But not all students will develop at the same pace and some may not have acquired either an understanding of the method or the skills at the end of a degree. In identifying those skills, this book is intended not to replace some elements of the undergraduate degree programme, but to complement it by helping you to appreciate the essential elements of physics that define it as a discipline and are required in its practice.

The scientific method in the undergraduate laboratory

You will probably question whether this process and the critical thinking skills that accompany it accurately reflect the nature of undergraduate experiments. The problem will already have been defined for you in the laboratory script, and to a great extent so will the experimental method. However, in order to do the experiment well you are still required to: gain knowledge (by background reading); understand the problem; apply that understanding in setting up the apparatus; collect data; analyse the data; synthesize, that is, draw conclusions from your work; and evaluate. Finally, of course, you will write a report on the experiment. Skill in writing reports is the subject of Chapter 7, but let me deal briefly with the immediate issue raised here.

▶ Report writing: a reader's perspective

It is important to realise that, just as you have applied these critical thinking skills to your experiment, your reader will apply similar critical thinking skills when reading your report. In particular, your reader will expect to understand what you set out to achieve, what you have done and why, and what you have achieved. Finally, your reader will expect to *learn* from the report. This means that your reader will appraise and evaluate the report and on this assessment will decide whether to believe what you write or to dismiss it. Your paper must reflect both the research process and the care you have taken.

You will be surprised to learn, perhaps, that even professional scientists miss this point. As a professional physicist I have refereed a number of papers for learned journals and a number of them fail to meet one important criterion; that is, if I were a reader interested in that particular field would I consider the paper a useful addition to the field or would I dismiss the paper as being not very useful to me. Why should I do the latter? Because as an interested reader I would want to have confidence that the results are meaningful. I would want to know whether the experiments have been well thought out, whether the underlying theory is sound, whether the experimental technique is sound, whether the results are credible, and ultimately, whether the conclusions are correct. I, as reader, have only the written manuscript by which to make these judgements. If the information is incomplete I am left in the position of recognising that even if the results are interesting and noteworthy I may not be able to attach any weight to them. The paper then has failed in its task of reporting a piece of scientific work to the community because the community, in exercising critical thinking towards the paper, rejects it as wanting.

The research process and scientific reporting

I emphasise again, it is essential that your paper reflects the critical thinking skills employed in the research process. We naturally place emphasis on the results. We may allow that negative results are important but we would not concede that a failed experiment has anything useful to say. We need to be very clear, therefore, about what is meant by a failed experiment and a useful negative result. This is an aspect of the last critical thinking skill – evaluation – described earlier and will be dealt with in more detail in Chapter 7. In essence a failed experiment is one that is badly thought out and badly conducted and fails to yield any useful information, but a useful negative result arises from a well-thought-out, well-conducted experiment where the method, and hence the results, are credible. If the experiment fails to demonstrate anything positive, for example the prediction of a theory, then at least it can be taken with confidence to be a fact.

The point to emphasise here is that it is not so much the result that matters as the experimental method itself, but at undergraduate level the distinction between a failed experiment and a negative result can be blurred. Under-graduate experiments are expected to 'work', that is, to yield a sensible value. We are often required to do experiments where we have a prior expectation of the result, for example measuring the refractive index of glass, the surface tension of water, the acceleration due to gravity. These things can be looked up in a book or on the internet. If the experimental result does not match the expectation, it is only natural to come to the conclusion that the result is not sensible and that the experiment has failed. This is not necessarily the case, however. The experiment may have been performed perfectly adequately. It is part of the skill in evaluating to assess the confidence in the result and to determine what, if anything, has gone wrong. To do otherwise is to shift the emphasis away from the process to the end result. Do this and you are no longer exercising critical thinking skills.

Evaluating your experiment

Consider further what is meant by the 'right' result. Tabulated data that appear in compendia of physical properties are themselves taken from experimental results. The data are sifted by specialists in the relevant fields who will make critical judgements as to whether a particular data set should be included. Some will but some will not. The same critical process that I have described in judging the quality of a report or paper is also applied by the specialists to the data under review. Data judged to constitute a 'standard' has often been measured very carefully but sometimes there may have been only one measurement in the literature which is then taken as the standard in the absence of any other. The result of your experiment may agree with the published data but ask yourself the question; 'Is this because the experiment has been well-designed and carefully executed or is it because the experiment has been badly designed and badly executed but by coincidence there is some agreement?' Should you ever find yourself in the position of being unable to answer this question then your experiment has been a waste of time. Concentration on the result, while entirely understandable, doesn't mean very much. It is much more important to concentrate on the process and to be able to evaluate the experiment.

▶ The scope of this book

You cannot learn undergraduate physics from this book, except for the specific topics described within. What you will learn, however, is a range of ideas, techniques and concepts that are essential to the physicist, including some specific mathematical and physical concepts. It cannot be any other way. In education parlance, physics is a 'vertical' subject. One thing builds upon another and the foundations for subsequent study must be laid at the outset. Thus it is not possible to talk about designing a good experiment without an appreciation of the purpose of experimentation. Data analysis cannot be discussed without knowledge of statistics, and theoretical physics cannot be discussed without reference to calculus and differentiation. All these topics must be described, therefore, but not in the detail you would find in a textbook, but briefly, though in sufficient detail for you to grasp the essence of the subject and its place in the edifice that is physics.

The mathematical skills will be approached from the point of view that maths is a tool to the physicist. The aim is to examine the role of mathematics in both experimentation and modelling, but especially the latter. This is the real art of the physicist; to be able to take a physical situation and describe it mathematically, and conversely to take a mathematical formulation and extract from it the physical implications. The mathematical manipulations in between these two extremes do not need to bear any specific relation to reality. They are simply part of the process and may be done by a mathematician just as easily as a physicist skilled in mathematical techniques.

In truth the range of mathematical techniques that a theoretical physicist will employ is enormous, and any attempt to describe the majority of them in detail would result in another textbook in mathematics, and there are plenty of those on the market. Rather, I will concentrate on what I consider to be the most essential mathematical techniques, such as differentiation and complex numbers, which find repeated application in our physical description of the world, and give examples to illustrate their use. The subject of modelling will be treated separately so that the practical issues involved in developing theoretical models can be explored without the distraction of having to consider the philosophy of the method.

Similarly with experimental techniques, the aim is not to describe individual experiments but to explore the processes involved in designing a good experiment. Inevitably this involves understanding the nature and sources of errors, quite apart from the calculation of the uncertainty on an experimentally measured parameter. Statistical analysis is described in Chapter 4, but as statistics finds such widespread use in physics the subject is approached through an exploration of statistical concepts. These, then, are the foundations of physics to be found in these pages. Where possible, the implications for your own learning and understanding are described, but remember, ultimately you are responsible for your own learning. The contents of this book will assist you in your aim to study and understand physics, but on its own it will not make you a physicist. That is up to you.

▶ How to approach the maths in this book

The mathematical topics described in this book are given with a view to showing how maths is related to physics. There are only two overtly mathematical chapters 4, and 6, of which the former is perhaps the hardest. It certainly looks the most daunting. Some of the expressions are quite long, but they have been developed from the basics. The maths throughout this book should be accessible to anyone familiar with algebraic manipulation (see Appendix 1 if you are not) and possessing rudimentary knowledge of calculus. It may take some effort but it should be relatively straightforward to follow.

Ideally you should understand every equation but that may not be possible. At least you may feel that it is not. Maths is as much about confidence as it is about ability. If you lose your confidence you will struggle, even if you are intellectually quite capable of following the arguments. Start at the basics. Do not expect simply to be able to read it as though it were a novel and to understand it. This will not happen unless you are already familiar with the maths. Try the following approach instead. Accept what is written as true and skim over the detailed maths so that you can pick up the gist of the argument. That is to say, look at the maths and understand it if you can but don't spend time over it. In some cases you will have no choice. Where equations are given without derivation there is nothing to understand. Just accept it. Balance the need to follow every detail as you go against the possibility that you will get bogged down and lose sight of the broader picture, which will probably happen if there is a mathematical step you find difficult to follow. Accept it, move on. Having read through and understood the essence, then return to the detailed maths so that you can appreciate the argument in full.

▶ Note

1 Robert Ehrlich, *American Journal of Physics*, vol. 70(1), 2002, pp. 24–9.

2 Mathematical Techniques in Experimental Physics

Skill in performing experiments is just as important in physics as skill in developing theory. It is easy, and quite common, to regard experimental physics as somehow the poor relation. In so far as theoretical physics often involves detailed mathematical arguments it is an impressive endeavour, but theoretical and experimental physics are best viewed as complementary necessities rather than alternative approaches. The theory tells us 'why', but the experiment tells us 'what'. A theory is, after all, an intellectual construct, and as such it is as prone to error as any logical argument, philosophical or theological. A well-designed experiment provides the 'facts' of the physical world, in so far as what we observe is what happens. If a discrepancy exists between the facts of the experiments and the predictions of a theory a debate about which is right and which is wrong will inevitably ensue, but if the experimental techniques are sound and there is confidence in the results, it points to errors in the theory. Without the definite 'facts' of experimental physics theoretical physics is ultimately without test and therefore without meaning.

An experiment may also be prone to error, of course. The skill of the experimentalist lies in part in eliminating as many errors as possible and quantifying those that cannot be eliminated. In order to do so, the experimentalist will draw upon a range of skills: knowledge of instrumentation and electronics, computational physics, use of statistics for an analysis of errors and accuracy, and so on. Many of these will be developed in subsequent chapters. The present chapter will concentrate on the essential mathematical techniques required of an experimentalist. Mathematics is a language, and like any language there are rules that govern its use. The purpose here is to define some basic mathematical rules and techniques that help both to formulate an experimental problem and to analyse the resulting data. The design of the experiment, that is how the measurements are to be performed, is dealt with in Chapter 3.

▶ Experiment in physics

Experimental physics is fundamentally about quantification. In some branches of science experimental methods have involved little more than cataloguing; this data from here is compared with that data from there, and so on. The truth is, of course, that science, or at least what we have come to know as science, is not a homogeneous activity. The methods of biology are different from those of chemistry, which are again different from those of physics. Physics at its most basic is concerned with understanding the natural world, and requires that we ask fundamental questions of how and why things happen. If something is pushed, does it move? How fast, and how far does it move? What causes it to stop? In other words, a wordy description is not much use; it is necessary to be able to express that description in mathematical terms so that the effects can be quantified.

This is not just a question of the development of mathematical theories. Any theory has to be backed up by experimentation and observation, and in many cases the experiment comes before the theory. If a theory has been developed it will of course inform the development of the experiment. Conversely, experimental observations can inform the development of a mathematical theory, a subject dealt with in greater detail in Chapter 5 where the relationship between mathematics and physics is explored, and in particular the use of mathematics as a form of logic to formulate arguments about the physical world. Basic to all these arguments, however, is the idea of a functional, or dependent, relationship between two or more physical quantities.

▶ Dependent relationships

A dependent relationship exists if: *property A depends on quality B.* The word 'quality' is used because *B* may be an action, such as a force, or it may be some inherent property such as the amount of charge or mass that a body has. This is a specific use of the word 'dependent' which is quite different from common usage. In everyday parlance you might say that whether you attend an open-air event, for example, will depend on the weather. This usage implies the notion of contingency and decision-making; a decision to do this or that is contingent upon such-and-such a condition being met. There are many examples of words used in physics that have very definite meanings quite different from the common usage. If we are not clear about the meaning of particular words then, not surprisingly, the way is open to all manner of confusion. Precision in thought is best achieved through precision in language and success in science is not possible without precision in thought.

To return to the notion of dependence, what physicists mean by this is perhaps best illustrated by means of an example. The story of Archimedes is well-known. Charged by the king to find a way of checking the purity of his gold to ensure that he was not being cheated, Archimedes eventually realised that the amount of water displaced by an object was related to, that is *depends* on, its density and hence he could determine the purity of the gold simply by immersing it in water. Such was his joy – his life depended on his being able to find a method – that he is reputed to have run down the street naked shouting the now famous word, 'Eureka'. Dependence, then, means that there is some relationship between the properties. If some condition of our experiment is altered and the result of a measurement is also altered then there is a dependent relationship between the two.

▶ Functions and variables

Experimental physics is all about testing dependent relationships. Sometimes an experiment will test a theory that predicts a specific relationship, but on other occasions the experiment will be performed to determine the dependent relationship. Mathematically the notion of dependence is expressed through the use of the function. To say that property A is dependent on quality B is also to say that A is a *function* of B, which is expressed as:

$$A = f(B) \qquad (2.1)$$

The notation $f()$ represents the function. B is known as the independent variable and A is the dependent variable, because in an experiment B is varied independently and A is measured. Therefore A depends on B.

Mathematical notation like this is a form of shorthand in which complicated ideas can be expressed concisely, but also very generally. A specific dependence – linear, quadratic, exponential for example – arises from a specific theory which may or may not be correct, but this does not stop us defining general relationships and developing mathematical arguments. For example, it is possible to derive the mathematical relationships between force, acceleration, velocity and distance so that it is possible to write the distance travelled in a given time by a body accelerated by a force. It is not possible to do the calculation, though, unless the force is specified along with the time over which it acts. This is a specific detail, but the mathematical argument is general, and the development of such general arguments helps to define the conditions under which an experiment must be conducted in order to generate meaningful results.

Returning to Archimedes it is possible to write, for example,

$$V_W = f(\rho) \tag{2.2}$$

where V_W is the volume of water displaced and ρ is the density of the metal. Explicitly this tells us that the volume of water displaced is a function of the density of the metal. However, it could be argued on purely logical grounds that there must be at least two bits of metal that are completely different but will nonetheless displace the same volume of water. This possibility can be eliminated if it is recognised that it is not just the density but also the mass that needs to be considered, and we should strictly write

$$V_W = f(m, \rho) \tag{2.3}$$

where m is the mass of the metal. V_W is now a function of two *dependent* variables. In other words, both variables can affect the outcome of the experiment. One of them must be held constant otherwise the measurement makes no sense. Clearly m must be fixed at some common value so the notation changes to

$$V_W = f(\rho)|_m \tag{2.4}$$

where the vertical line indicates that m is to be fixed. Fixing m effectively reduces this to a function of one variable.

Symbols and notation

One of the confusing notions of the function is the choice of symbol. By convention f is normally used to express a functional relationship, but any letter can be used. For example, instead of expressing the volume of water displaced, the mass of the water displaced may be written:

$$m_W = g(\rho)|_m \tag{2.5}$$

The two functions, f and g, are not the same because the volume of water is not the same as the mass of water, so the two cannot have the same functional relationship to a third variable. The mass of water can be related to its volume by the density, and experience tells us that the density of water varies with, that is, is *dependent* on, temperature and so we write

$\rho(T)$. Hence, changing the dependent variable can cause another independent variable – in this case T, the temperature – to be introduced. This temperature dependence is incorporated into equation (2.5) by writing

$$m_w = g(\rho, T)|_m \tag{2.6}$$

Experimental constraints
Equation (2.6) can be further simplified by putting T constant, that is:

$$m_w = g(\rho)|_{m,T} \tag{2.7}$$

To re-emphasise, equation (2.7) is a shorthand for a lengthy verbal argument that says the mass of water displaced by a piece of metal depends on:

- the density of the metal,
- the mass of the metal, and
- the temperature at which the experiment is performed,

so for the experiment to yield anything useful both the temperature and the mass of the metal must be kept constant. This doesn't say anything about whether this is the *best* experiment to perform (it might be easier to measure the volume of water displaced, for example), but it tells us in a very neat and concise way what conditions must be imposed if this experiment were to be performed.

▶ Data, functions and graphs

Suppose that the experiment has been performed so that a set of figures exists that relates the mass of water displaced to the density of the metal. We now want to know exactly what this relationship is. More generally, given a set of measurements of a at set values of b, that is, (a_1, b_1), (a_2, b_2), $(a_3, b_3) \ldots (a_n, b_n)$ and so on, we want to know how to calculate a at any given b. To a certain extent we can refer back to the data if we want to know specific values of a corresponding to specific values of b. We know that when $b = b_1$ for example then $a = a_1$, but what if we want to know the value of a if b lies between two of the values chosen for the experiment? We can't necessarily return to the experiment and take a new measurement, though of course there may be circumstances where that is possible. What is required is a method of calculating a given any value of b we choose. Mathematically this can be done by expressing the data in the form $a = f(b)$, where the exact nature of f is specified.

The first step is to plot a graph. A graph allows for visualisation of the data and is far superior to a table of values. In general a table of numbers is very hard to interpret, except in special circumstances. Consider the following simple exercise (Table 2.1) with two variables, say current (I) and voltage (V), where I is the dependent variable.

It is easy to see that

$$I = 2V. \tag{2.8}$$

However, experimental data is rarely this perfect. Consider instead the set of data shown in Table 2.2. The functional form is not immediately obvious but it is obvious that the data is different. Not only are the values different, but in Table 2.1 the values of I are exact, whereas in Table 2.2 the data are quoted to one decimal place. It is sufficient for the purposes of this case to note that the numbers are slightly different from those in Table 2.1 and the question of the cause of the 'scatter' of the data will not be addressed. In any real experiment, of course, the question arises, 'Is the variation caused by experimental error or is the data systematically different?' and this is answered in part by looking at the accuracy of the measurement and the

Table 2.1 *Idealised V–I data*

V	I
1	2
2	4
3	6
4	8
5	10
6	12

Table 2.2 *Similar data to Table 2.1*

V	I
1	2.3
2	3.7
3	6.0
4	8.3
5	9.6
6	12.4

Table 2.3 *A slightly different set of data from both Tables 2.1 and 2.2*

V	I
1	2.5
2	3.8
3	5.5
4	7.5
5	9.9
6	12.6

confidence that can be placed on any one result. It is not a question that can be easily answered simply by looking at a table of data, though.

Consider a third set of data as in Table 2.3.

This is also different from Table 2.1, and also Table 2.2. In respect of the latter the deviation in some cases is slightly larger, but is there any systematic difference between the different tables? It is impossible to tell on the basis of the tables alone.

Plot a graph, however, and the situation immediately becomes clear (Figure 2.1). In (a) the data from Table 2.1 only are plotted and they all lie on a straight line. In (b) data from Table 2.2 are plotted as symbols and, as can be seen, the points are scattered about the straight line in no particular order. In (c), however, the data from Table 2.3 are plotted similarly but there appears to be a systematic deviation from the straight line. Instead of a seemingly random scatter, the points seem to be arranged in a curve. If this was the result of an experiment the first suspicion might be that a slight curve would better describe the data, meaning that Ohm's law is not obeyed.

A word of caution at this point; interpreting experimental data is often not as simple as this. In a real experiment, other factors must be considered before any firm conclusions on the nature of the curve can be reached. Errors have already been mentioned, and one effect of errors is to introduce some variation into the data so that it is distributed about some 'perfect' curve as represented by Table 2.2 and Figure 2.1(b). The task of data analysis is to determine this 'perfect' curve. In Figure 2.1(b) we would have no difficulty in accepting that this perfect curve is a straight line, but in Figure 2.1(c) we would have to decide between a straight line and a curve, and with it whether Ohm's law is obeyed. There are physical consequences to our decision so it is important therefore to be clear about how we arrive

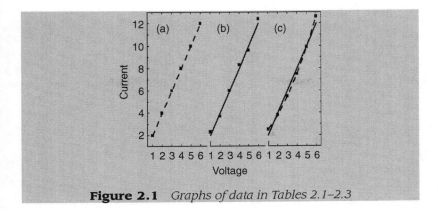

Figure 2.1 *Graphs of data in Tables 2.1–2.3*

at such conclusions. In this particular case the question is somewhat artificial because I have chosen this data to lie on a curve specifically to show how much better a graph illustrates what is happening than a table. If it were real experimental data though, it might well be scattered much as the data in Figure 2.1(b) is scattered about its curve so the deviation from the straight line is not so obvious. Detailed analysis of experimental data has to be based on a sound understanding of sources of error and of statistical techniques, and is dealt with in more detail in Chapter 4.

▶ Functional forms and mathematical relationships

This has not answered the original question of how to go about finding the mathematical function – the 'perfect' curve – that describes the data. In general this is not an easy problem and some prior familiarity with functions is necessary, but there are some cases that are easier than others. In the case of the straight line, which is one of the simplest mathematical functions, the equation is easy to derive. Where the deviations from the 'perfect' case are small or non-existent, the straight line through all the data can be drawn by eye. In the case where the scatter may be large, drawing a graph by eye is somewhat haphazard, and will not necessarily lead to the best possible description of the data. In this case the best straight line ought really to be calculated according to the method of least squares described in Chapter 4. However, in the absence of the necessary mathematical tools a straight line can be drawn as a good estimate of the idealisation of the data.

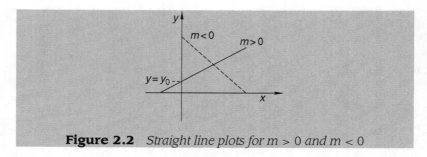

Figure 2.2 *Straight line plots for m > 0 and m < 0*

Straight line plots

Suppose, then, that a line has been drawn as in Figure 2.2. The form of the mathematical function is derived in a simple manner. There are two key parameters that specify a straight line; the slope and the intercept. The intercept is defined as the value of y when $x = 0$, also written also as $y(0)$. Therefore, let

$$c = y(0) \tag{2.9}$$

be the intercept. The slope is defined as the rate of change of y with x. For any two values of $y_1 = y(x_1)$ and $y_2 = y(x_2)$, where $x_2 > x_1$ then the slope is defined as,

$$m = \frac{y(x_2) - y(x_1)}{x_2 - x_1} \tag{2.10}$$

and the equation of the straight line is

$$y = mx + c \tag{2.11}$$

For the data in Table 2.1, $m = 2$ and $c = 0$. This leads to the same result as before, viz. that $y = 2x$.

A number of combinations of m and c are possible. The slope can be positive, in which case an increase in the independent variable produces an increase in the dependent variable, or it can be negative, so that an increase in the independent variable produces a decrease in the dependent variable. The intercept can be zero, as above, or positive or negative, giving six possible forms of the straight line graph.

Physical examples of straight line plots

Concentrating on the case where $m > 0$, the various values of the intercept represent real physical phenomena:

- An intercept $c > 0$ usually represents an initial or 'built-in' value.
- For $c < 0$, the form of the straight line usually implies some sort of threshold.

Example 2.1

Consider the well-known kinematic equation of motion for a body subject to a constant acceleration a,

$$v = u + at \qquad (2.12)$$

where the variable v (strictly the function $v(t)$, represents the velocity as a function of time t. u is the initial velocity, that is in the notation of the function $u = v(0)$. By comparison with the equation for the straight line, $u = c$. At time $t = 0$ the velocity is u and the velocity increases linearly with time. The change in velocity in time t is given by at and this is added to the initial, or in-built, value u.

Naturally, the question arises, 'What about values of the velocity $v < u$?' These are only defined for negative t, or $t < 0$. In reality this is not something of concern. The definition of $t = 0$ is entirely artificial corresponding to the point when the clock is started. If it made sense to measure earlier events the clock would be started earlier, if that were possible. It follows then, that we have no real knowledge of events for $t < 0$, so strictly the line isn't defined for these times. It doesn't mean to say that the velocity cannot be *extrapolated* back to $t < 0$, as the term is called. Experimentally, our data may be limited in range but the mathematical form of the function, being ideal, is not so we can easily calculate what value of v we might expect at $t < 0$, *but that doesn't mean to say that this corresponds to any physical reality*.

Example 2.2

Suppose we wanted to use a high-power pulsed laser to melt the surface of a metal. A pulsed laser is ideally suited to this task because energy is dumped into the surface very quickly causing considerable heating. We can imagine that the amount of material melted increases with the energy of the laser pulse. This is in fact the basis of laser welding. However, if there is insufficient energy in the laser pulse the surface simply heats up but does not melt, so there is a *threshold* energy for melting the surface. In a plot of the depth of melting against the laser energy the first few points would lie on the x-axis before increasing. In reality the dependence may not be linear but let's suppose for the sake of argument that it is (Figure 2.3). This corresponds to a straight line of positive slope and negative intercept (extrapolated back to $x = 0$). Mathematically it is

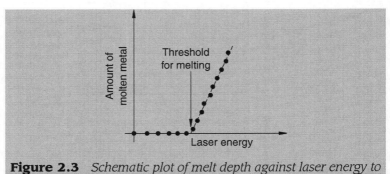

Figure 2.3 *Schematic plot of melt depth against laser energy to illustrate the concept of a threshold*

possible to extrapolate the line back to zero laser energy and negative melt depths, but physically it makes no sense. The data is not defined, but more importantly neither is the concept of a negative melt depth.

Proportionality

The straight line with an intercept of zero expresses the notion of *proportionality*. If two properties have a proportional relationship then a change in one will cause an equivalent change in another. For example, if the independent variable is doubled, then the dependent variable is doubled. Proportional relationships are very common in physics. For example, Newton's second law has it that, '*A body subject to a constant force experiences an acceleration in the direction of the force which is* **proportional** *to the magnitude of the force* and **inversely proportional** *to the mass of the body.*' Proportionality is expressed using the symbol $\propto$, so mathematically Newton's second law is written:

$$a \propto \frac{F}{m} \tag{2.13}$$

where F is the force and m is the mass. 'Inversely proportional' means 'proportional to the inverse', so the mass appears in the denominator.

Similarly, Ohm's law is often expressed as a proportionality; the voltage V dropped across a wire carrying a current I is proportional to the current. Mathematically,

$$V \propto I \tag{2.14}$$

Ohm's law demonstrates very neatly the relationship between the function and the experiment. As expressed above, this law implies very clearly that a current is supplied and a voltage measured. The current is the independent variable, and the voltage the dependent variable. This may indeed have been the experiment that Ohm performed originally, but today it is very unlikely that an experiment to verify Ohm's law would be performed in this manner. As power supplies are very common it is much more likely that a voltage would be supplied and a current measured. The current therefore becomes the dependent variable and

$$I \propto V \tag{2.15}$$

A proportional relationship does not, however, imply a one-to-one relationship. In certain special cases this might apply, as with Newton's second law:

$$a = \frac{F}{m} \tag{2.16}$$

In order to convert a proportionality to an equality, a multiplying factor is required, called 'the constant of proportionality'. Thus if A is proportional to B we write

$$A = k \cdot B \tag{2.17}$$

where k is the constant of proportionality. For one-to-one relationships, $k = 1$. For Ohm's law we write

$$V = R \cdot I \tag{2.18}$$

where the constant of proportionality is a property of the material known as the resistance, R. In the inverse relationship,

$$I = \frac{V}{R} \tag{2.19}$$

▶ Power-law dependencies

The power law is very common in physics. Expressed as a proportionality, the power law of the dependence function $y = f(x)$ is written:

$$y \propto x^n \tag{2.20}$$

where n is some value other than one (Figure 2b). If $n = 1$, of course, the relationship is linear, as already described. For values of $0 < n < 1$ the dependence is said to be sub-linear; y increases with x but the rate of increase decreases with x. For $n > 1$ the dependence is super-linear; y increases as x increases, but the rate of increase increases with increasing x. For $n < 0$, the behaviour is asymptotic for large enough values of x. That is to say, y will decrease as x increases but at no stage will y become

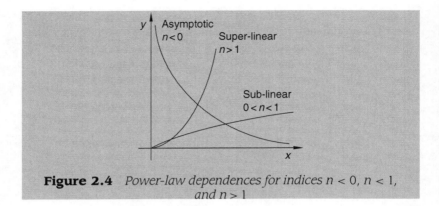

Figure 2.4 *Power-law dependences for indices $n < 0$, $n < 1$, and $n > 1$*

negative. The most that will happen is that y will approach zero but it never quite reaches it.

Integer values of n correspond to special cases; parabolic for $n = 2$ and cubic for $n = 3$. Well-known examples of the former include the dependence of kinetic energy, K, on velocity, v.

$$K = 1/2mv^2 \tag{2.21}$$

and for the latter, the relationship between volume, V, and the length, l, of the side of a cube:

$$V = l^3 \tag{2.22}$$

For an example of the fourth power look no further than Stefan's law and the Stefan–Boltzmann constant. A hot body loses energy to the environment by emitting electromagnetic radiation. As the temperature of the body increases the wavelength of the emitted radiation decreases, hence at about 670°C the body appears red, then yellow, and eventually white hot at over 1000°C. The amount of power P emitted from unit area of the surface is proportional to the fourth power of the absolute temperature, and the constant of proportionality is the Stefan–Boltzmann constant, σ, that is:

$$P = \sigma T^4 \tag{2.23}$$

Polynomial functions

A polynomial is a function which contains terms in increasing powers. A polynomial of order n has the form:

$$y = a_0 + a_1 x + a_2 x^2 + a_3 x^3 + \ldots + a_n x^n \tag{2.24}$$

The order of the polynomial is equal to the highest power in the series. For example, a polynomial of order 1 is the linear equation, that of order 2 is a quadratic equation, and that of order 3 is a cubic equation. The coefficients $a_m, m = 0$ to n, can be negative, zero or positive. If the coefficients of all but the highest term are zero the polynomial becomes the power law of the last section.

Theoretically physics doesn't make much use of the polynomial, except in its lowest order forms of quadratic and cubic. An example is the kinematic equation describing the distance $S(a, t)$ travelled by a body in a time t subject to a constant acceleration a and having an initial velocity u.

$$S = ut + \frac{1}{2}at^2 \tag{2.25}$$

which is quadratic in t. Experimentally, however, the polynomial function is very useful. When trying to express experimental data in the form of a function, the polynomial is often most appropriate. Referring back to the case of the thermocouple, the measurement of the voltage as a function of temperature will yield a large amount of data which will be expressible in some mathematical form. The mathematical form does not have to possess any physical significance. That is, it does not need to be based on any particular theory, it simply needs to be in such a form that the voltage can be substituted into an equation so that the temperature can be calculated. Polynomial functions are extremely useful for this sort of purpose.

Exponential functions

The exponential function is perhaps the most common in physics. All sorts of natural phenomena are described mathematically by a function of the sort

$$Y \propto e^\alpha \tag{2.26}$$

where α may be positive or negative. In particular, the exponential is almost always the most appropriate mathematical form of a decay. Examples include:

- Radioactive decay; the number of active nuclei N decays exponentially with time t:

$$N = N_0 e^{-\alpha t} \tag{2.27}$$

   where N_0 and α are constants.
- Optical absorption; the intensity of light I inside an absorbing medium decays with distance x:

$$I = I_0 e^{-\alpha x} \tag{2.28}$$

   where, again, I_0 and α are constants.

- A damped harmonic oscillation; the amplitude of a harmonic oscillator decays with time according to:

$$A = A_0 e^{-\alpha t} \tag{2.29}$$

In a perfect oscillation y will alternate between $+A$ and $-A$ indefinitely according to equation (2.30), but in reality no vibration lasts forever, and the amplitude A decays to zero as the oscillation stops:

$$y = A\sin(\omega t) \tag{2.30}$$

It will be immediately apparent that the same basic mathematical description applies to a variety of different physical phenomena, and as you develop your knowledge in physics you will come to realise just how universal are many of the concepts. These three examples are described in more detail in Chapter 6.

The exponential function is also used to describe processes which are said to be 'activated'. Conduction in semiconductors and rates of chemical reactions are two examples of activated processes described by,

$$y = Ae^{-\beta/T} \tag{2.31}$$

where A and β are constants and T, is the absolute temperature. Activated processes are fundamentally probabilistic, that is to say the probability of an event y occurring increases with temperature. Thus in conduction in a semiconductor device it is the probability of finding an electron available for conduction that increases, and in a chemical reaction the probability that the reaction will take place within a certain time increases with temperature. Again, this is an example of a universal phenomenon. Even the decay phenomena described previously could be couched in the language of probability; the probability of finding an active nucleus at a time t, or the probability of finding a photon at a depth x. These theoretical expressions are not always derived using probability, but the fact that such phenomena can be interpreted in this manner is illustrative of the universal nature of much of physics.

Logarithmic plots

The non-linear functions such as the power law and the exponential function are not always sensibly plotted on the conventional graph with linear axes. Consider the fourth power law for the power radiated as a function of temperature. A body at 600°C will radiate nearly 100 times more power than a body at room temperature, so plotting this on a normal graph will result in a large number of points appearing to be zero or at least very small.

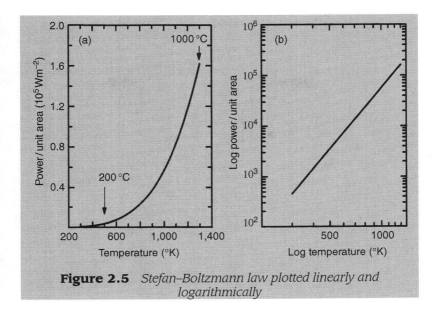

Figure 2.5 *Stefan–Boltzmann law plotted linearly and logarithmically*

The way around this is to use a logarithmic scale for one or both of the axes (Figure 2.5).

It can be shown (see Appendix 2) that for a power-law relationship of the form $y = ax^n$, where a is a coefficient,

$$\log(y) = \log(a) + n\log(x) \tag{2.32}$$

which is now expressed in the form of a straight line with $\log(y)$ being plotted against $\log(x)$. The slope is n and the intercept at $\log(x) = 0$, that is $x = 1$, is $\log(a)$. This type of plot is referred to as a 'log–log' plot, as logarithms of both axes are taken. For exponential function the use of Naperian logs results in a 'log-linear' plot. If

$$Y = a.e^{bx} \tag{2.33}$$

then

$$\ln(y) = \ln(a) + bx \tag{2.34}$$

which is a straight line if $\ln(y)$ is plotted against x, with slope b and intercept $\ln(a)$.

▶ Transformations: data manipulation and algebra

The use of the logarithm is a transformation of the data from one form to another. The purpose can be either to allow data spanning several orders of magnitude to be plotted sensibly or to allow a straight line function to be plotted from which the parameters of the function, that is the exponents and coefficients, can be determined. As described, the straight line is the easiest function to use. Deviations are immediately obvious and the slope and intercept are easily calculated. Non-linear behaviour can also be identified readily but the form of the function will not be so apparent from visual inspection. Where possible, transforming the data to arrange it in such a form that an easily recognisable function can be plotted is a useful experimental technique.

For power-law behaviour of a known form, for example a square law or cubic law, plotting the square or the cube of the parameter is sufficient. This will yield the desired straight line. If either a power law or an exponential is suspected take logarithms and plot according to equations (2.32) or (2.34). For polynomial functions the transformation is not so clear. Consider the kinematic equation:

$$S = ut + \frac{1}{2}at^2 \tag{2.35}$$

Plotting a measurement of S against t will not yield a straight line because of the super-linear t^2 term. However, nor will plotting S against t^2 yield a straight line because the term in t will be sub-linear. In this particular case it is possible to manipulate this function algebraicly and to divide through by t to give

$$\frac{S}{t} = u + \frac{1}{2}at \tag{2.36}$$

so that a plot of S/t against t yields the desired straight line of slope $1/2a$ and intercept u. Algebraic manipulations of this sort often enormously simplify the business of interpreting experimental data. If you are not sure about the techniques of algebraic manipulation they are described in greater detail in Appendix 1.

▶ Curve-fitting

Where such transformations are not possible, other methods of generating the function and deriving the parameters are needed. One very common method is curve-fitting, which can be done with most commercial scientific graph-plotting packages. Polynomials up to order 9 are very common and usually there is some facility for user-defined functions. Curve-fitting is particularly useful in the case of, say, a calibration, where some effect, such as a deflection on a meter, has to be related to the property causing the deflection. Here it is necessary only to find a mathematical function that describes the data. There is no physical meaning attached to the parameters so it is not particularly important what constraints are placed upon them. However, if the purpose of fitting a curve is to derive some physical meaning it is necessary to be clear about the limitations of the method.

In most cases knowledge about the detailed mathematics of the curve-fitting algorithm is not necessary. However, it is necessary to appreciate in general terms what the computer is trying to do in order to understand the possible deficiencies in the outcome. All curve-fitting routines work by calculating a curve of a specified mathematical form and comparing it with the data according to some criterion. The parameters of the function, some of which may be fixed at the outset, are then varied by the computer and a new curve calculated. If the new curve is better, according to the 'goodness of fit' criteria, these new parameters are accepted. When no further improvement is possible the programme ends and the final values of the parameters define the mathematical function. In the course of fitting a curve hundreds, or perhaps thousands, of trial curves will be calculated before the final parameters are chosen. Different curve-fitting algorithms vary in the method of changing the parameters and accepting the changes, all of which affects the speed and efficiency of the programme, but is not important for our purposes.

The criterion by which one curve is judged to be better than another is most commonly chosen to be the 'sum of the squares'. In a plot of y against x, for example, it is possible at each value of x to calculate the difference between the experimentally measured y and the value of y calculated according to the mathematical function. Clearly, the smaller this difference the better the function, so by making such a comparison over the whole range of x it is possible to arrive at a suitable criterion for a 'good' fit. The discrepancy between the calculated and experimental points is minimised by summing the *squares* of the differences for all pairs of (x, y) rather than summing the differences. In Figure 2.6, for example, the negative differences in the centre of the graph are offset to a great extent by the positive differences on the left and the right, but the square of these difference will always be positive.

Minimising the sum-of-squares in this manner is a straightforward computational operation. There is no reference to any physics; the computer is

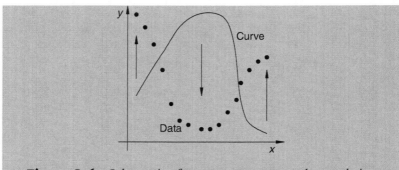

Figure 2.6 *Schematic of a computer-generated curve being compared with data. Upward arrows correspond to positive differences*

incapable of making judgements based on physical arguments. *It is up to you, the user, to make such arguments and these will be reflected in your choice of function.* For example, suppose you had a set of experimental data you believed followed a parabolic law. You can plot the square of the data, use a log–log plot, or try to fit a parabola, or possibly a more general power-law function, to a plot of the dependent variable against the independent variable. Not all of these methods are guaranteed to give you the same answer. They will be close, but how close? Is the discrepancy acceptable? Fitting a parabola where the power is fixed at 2 but the coefficient is allowed to vary will give you a particular solution with a particular sum-of-squares, but fitting a power law with the power as a variable in addition to the coefficient might result in a 'better' fit where the sum-of-squares is smaller but the power deviates slightly from 2, for example 1.95. Would you confidently assert that the data does not follow a parabolic behaviour? In fitting a power law the computer is not constrained to make the power an integer. If it can reduce the sum-of-squares by changing one of the variables, it will do so, but the result of such a change does not represent any physical reality. It is purely a mathematical operation and *if there is a good physical argument as to why the power should be 2, it is up to you to decide that this is the case and constrain the fit.*

Curve-fitting routines are very useful, but they have to be used with caution. Compare the operation of fitting a straight line to a log–log plot to fitting a power law to some data, for example in Figure 2.5. Ideally the fitted power law should give the same straight line as the log–log fit after transformation, but in fact the two methods are radically different. The data ranges over two orders of magnitude, from $\sim 10^3$ to $\sim 10^5$. A difference of 1 per cent between the data and the calculated curve at the highest temperatures will make the same contribution to the sum-of-squares as a 100 per cent difference at the lowest temperatures. Therefore in minimising the sum-of-squares the most weight will be given to the largest values, but in fitting to a log–log plot, where the data ranges from ~3 to ~5 all the data will be given a similar weight. The result will be that different values of key parameters are returned from fitting data in one form compared with another. When attempting to extract physically significant data from a curve fitting exercise it is imperative that the fitting process itself be informed by the physics of the problem.

▶ Differentiation and integration

The final topic of this chapter concerns the calculus of experimental data. Having derived a function to describe the data it may be necessary to perform other operations on the data, such as differentiation and integration. The differentiation of y with respect x is written as dy/dx, sometimes written as,

$$\frac{dy}{dx} = f'(x) \tag{2.37}$$

In the notation of the function a differential is represented by a superscripted dash, which can be done as many times as is necessary. For the second differential:

$$\frac{d^2y}{dx^2} = f''(x) \tag{2.38}$$

For a power law function of the form

$$y = ax^n \tag{2.39}$$

the differential is:

$$\frac{dy}{dx} = nax^{n-1} \tag{2.40}$$

As shown in Appendix 3, this is exact for a power-law dependence. If the function contains additive terms in different powers each term is differentiated separately and the total differential is the sum of all the individual terms. This is very useful for other functions, such as sine, cosine, exponential, or logarithm, which can be expressed as series (see Appendix 3).

There are three principal reasons for differentiating experimental data: calculating the slope of a graph to derive a quantity, calculating an error, and locating a maximum or minimum in the data.

1 *Calculating the slope.* In the case of a straight line plot the slope of the graph is easily derived and applies to all the data. For non-linear data, however, the slope has to be calculated at every point. Such a situation might arise for example, when trying to calculate the velocity of an object. Direct measurement of velocity is quite difficult; it is much easier to measure the distance travelled, S, at particular points in time and to differentiate the data according to (2.41):

$$v(t) = \frac{dS}{dt} \tag{2.41}$$

Differentiating experimental data directly is not likely to lead to accurate results, however. Consider, for example, Figures 2.1(b) and 2.1(c); the slope can be found at any point by taking the differences between successive data points, that is:

$$m = \frac{y(x_2) - y(x_1)}{x_2 - x_1} \tag{2.42}$$

Such a gradient is an average for the interval between x_1 and x_2, which is not very useful if the data is a curve as in Figure 2.1(c). Moreover, if the data is subject to scatter, as in Figure 2.1(b), the gradient will vary along the graph, so the method is not very accurate. It is much better to fit a function to the data and differentiate the function directly. In effect, this is done in Figure 2.1(b), where a straight line $I = 2V$ represents the 'ideal' curve to the data and the slope of the ideal curve is therefore 2. The function chosen to represent the data simply has to reproduce the data accurately over the range for which the differential is required.

2 *Calculating an error.* The error on a measurement is simply the confidence we place on the reproducibility of the result (see Chapters 3 and 4). We may measure x to an accuracy of $\pm\Delta x$, but how then do we find the corresponding accuracy in, say, y which is derived from x. We do it by differentiation, that is making the assumption that over the range of the error in x the function $y = f(x)$ is linear (Figure 2.7). Then:

$$\Delta y = \frac{dy}{dx} \cdot \Delta x \tag{2.43}$$

3 *Locating a maximum or minimum.* Figure A3.1 of Appendix 3 shows that the differential becomes zero at either a maximum or a minimum in the data. Differentiating the experimental data by fitting a function, most commonly a polynomial, to the data therefore constitutes an exact method of locating these points.

Integration is the reverse of differentiation. Therefore, if dy/dx is integrated this must return the function y. Just as differentiation has the mathematical meaning of calculating the gradient, integration also has a

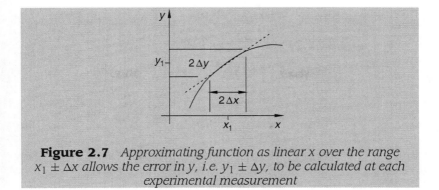

Figure 2.7 *Approximating function as linear x over the range $x_1 \pm \Delta x$ allows the error in y, i.e. $y_1 \pm \Delta y$, to be calculated at each experimental measurement*

mathematical interpretation, equivalent to finding the area under the graph. Integration of experimental data can be performed in an analogous manner to differentiation, that is by fitting some function to the data to be integrated.

▶ Summary

The focus of this chapter has been on the mathematics relevant to experimentation. The treatment is by no means exhaustive; this material is not intended to replace other mathematics texts, where some of the mathematical concepts are explained in much greater detail. Rather, the intention is to emphasise the role of these mathematical techniques in experimental physics in order to help you develop your skills in the context of your own learning and laboratory work. The mathematics is simple, and concentrates primarily on dependent relationships and their expression in the mathematical language of the function. The experimental problem can be clearly defined in terms of dependent and independent variables, and the parameters that need to be held constant. Awareness of simple mathematical functions, together with graph-plotting, algebraic manipulation and transformations of experimental data, and curve-fitting are all important for exploring functional relationships within experimental data. Differentiation and integration are shown to be valuable tools to the experimentalist not only for defining maxima and minima in the data, but also for deriving other properties, such as errors, from measured data.

3 Experimental Physics: Designing the Experiment

At undergraduate level, experiments are much less demanding than at postgraduate or professional level. The problem is already set out in a laboratory script as is the method of tackling it. One of the key purposes of experimental work at this level is to acquaint you with equipment and how it may be used, as well as some of the principles of experimentation, through practise. At some point in your undergraduate career, most likely towards your final year, it will be necessary to move beyond this to designing and implementing your own experiments and by the time you graduate you will be expected to have achieved a certain competence. The undergraduate laboratory should provide an environment in which you can learn the essential skills involved in designing an experiment.

Designing an experiment requires an understanding of many different things. Not least among them are:

- knowledge of the physics of the problem;
- an understanding of the equipment and what may sensibly be measured;
- an understanding of the sources of errors; and
- an understanding of the treatment of the data and errors in order to arrive at a final result.

In this chapter, emphasis is placed on the sources of errors, on the equipment (or instrumentation), and on the knowledge of the physics. Treatment of the data, also called 'data analysis', is left to another chapter. This is an inherently mathematical subject, and consideration of it alongside the physics of the experiment and the origin of experimental errors may make understanding the subject of this chapter more, rather than less, difficult.

▶ Dependent relationships and experimental choices

Designing an experiment requires that you choose which measurements are to be made. The previous chapter showed how an understanding of mathematical relationships can help to define the problem by showing clearly which variables are important and which are not, but it doesn't provide the means by which to decide how the experiment is to be conducted. For example, consider the Archimedean experiment described in the last chapter. Consideration of the dependent relationships has already defined that the mass of the metal must be kept constant, but quite possibly the temperature must also be fixed if the mass of the water displaced rather than the volume is measured. What is not clear, though, is whether it is better to measure the mass or the volume, or how to go about doing either.

In fact, a number of things could be done by way of experimentation, including:

- Compare the volume of water displaced by two or more pieces of different metals of the same weight, as already discussed.
- Measure the mass of water displaced by two or more pieces of different metals of the same weight, as already discussed.
- Systematically investigate how much water is displaced as a function of the weight so that we could distinguish between metals.
- Systematically investigate the amount of water displaced by gold alloyed with known amounts of impurity.
- Start to develop theoretical models of alloying and investigate any predictions of the model.

These are all experiments of one sort or another. Some have a theoretical component so the purpose of the measurements will be to test specific theoretical predictions. Others are more simple in their aims, which are to find experimentally some relationships between volume (or mass) of water and the density of the metal. Some criterion for deciding upon the course of experimental action is needed.

▶ The problem of measurement

Whatever experiment is decided upon, the problem of measurement is immediately encountered. Some measurements are easier than others and will therefore be favoured, which might ultimately decide which particular experiment is preferred. Before any decisions can be taken about the experimental arrangement it is necessary to decide what it is that should be measured. Will it be:

- An absolute value?
- A comparative measurement looking at differences or small changes?
- Or a null measurement, where the total effect is zero?

The answer to these questions requires an awareness of the capabilities and limitations of particular experimental techniques, especially instrumentation. Of particular interest is:

- An awareness of the sensitivity and resolution of the equipment.
- A knowledge of the likely sources of errors.

Some of the possible types of measurement are described below.

Absolute measurements
An absolute measurement would record the actual volume or mass of water displaced. Such measurements are the most difficult to make; not only do you have to be certain that your measurement instrumentation is capable of detecting the changes of interest, you also need to be certain that your measurement relates to a physical quantity. All possible sources of errors need to be identified and eliminated where possible.

Absolute measurements can require the use of imagination in order to be able to get at the quantity required. Consider the Archimedian options described above. Experimentally it is almost impossible to measure the mass of displaced water directly. No matter how hard you think, it will be difficult to devise a method of extracting this water in order to be able to put it directly onto scales. This does not mean that measurement of the mass is impossible. By no means! I shall describe a system to do just this, but it requires a thorough knowledge of the physics of the problem, as well as an appreciation of a particular arrangement that I shall call the null configuration and describe under that heading below.

Comparative measurements
In comparative measurements, one quantity is directly compared with another. Such measurements are extremely useful because there is no need to calibrate instruments and one of the major sources of systematic errors is eliminated. Indeed, in some cases it is possible to eliminate all experimental errors. The measurements, being comparative, are subject to the same errors, so if a difference between two otherwise identical experimental arrangements is detected it must be real, no matter what the source or magnitude of the errors is in actuality.

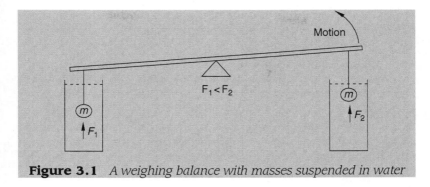

Figure 3.1 *A weighing balance with masses suspended in water*

In relation to the experimental options faced by the modern day Archimedean, for example, comparing the volume of water displaced by two or more pieces of metal constitutes just such a measurement. It is not necessary to know the actual amount of water displaced, which is an absolute measurement, only to be able to detect a difference. A system to do this might comprise a weighing balance with two pieces of metal of identical mass suspended in water (Figure 3.1). Archimedes' principle states that the weight of an object in water appears to be reduced by an amount equal to the weight of water displaced. The displaced water actually exerts an upward force (F_1 and F_2), and the effect of gravity is countered to some extent. Therefore the metal piece that displaces the least volume will be subject to the smaller upward force, and hence the balance will tip in the direction of the 'heaviest' metal; that is, the one that displaces the smallest mass of water.

You may be confused as to whether mass or volume is being detected in this arrangement. The answer is that either is applicable, as one transforms to the other via a simple mathematical relationship. It simply depends on your point of view. The weight – or *mass* – of the metal is being measured, but this is mediated by the *volume* of water displaced.

Null measurements

The essential principle of a null measurement system has two equal elements or arms. In the case of the balance illustrated above the arms are physical. However, in other systems the term may refer to an optical path or a resistive element, for example. The change to be measured is introduced into one element and the other adjusted until equality is re-established. The adjustment constitutes the measurement, but as the change is always zero the reference point in the measurement remains the same and the problem of deciding upon the magnitude of the change is eliminated. As before, many of the errors that can affect the measurement are eliminated because the same errors are found in both arms. The null configuration has the advantage that it is often easier to detect when something goes to zero than to measure precise changes.

Instead of allowing the scales to tip in order to detect a difference between two metals immersed in water, it would be possible to add weights on one end of the arm until the balance is re-established. A weighing balance is itself a null instrument; the weights on both arms balance each other out to give zero, or null, deflection. Used this way, the measurement changes from comparative to absolute if the weights and balance are calibrated correctly.

What measurement?

Not the name of a popular consumer magazine for physicists, but a difficult question that needs to be answered. As the preceding discussion has illustrated, there are numerous experimental arrangements possible for even a seemingly simple phenomenon such as Archimedes' principle. No doubt there are things that could be done that have not been mentioned, or possibly variations on some of the themes already discussed. In designing an experiment, you will have to decide exactly what it is you wish to achieve, why this in particular, and set about considering the physics of the measurement and the possible sources of error. The first two constitute the first stage in the research process described in Chapter 1 and utilise the first two critical thinking skills, whilst the last is part of the second stage, that of planning the research.

▶ Errors and accuracy in measurement

Two measurement systems have just been described in which many of the experimental errors do not have to be considered because the particular experimental arrangement allows for a common effect. In the other experiments, though, the problem of making accurate and precise measurements has to be faced, particularly if the predictions of a theory are to be tested. Precision and accuracy are the watchwords of the experimental physicist. Galileo, commonly recognised as the first experimental physicist, described physics as 'book-keeping'. Galileo did not have the advantages that we have today; the statistical methods developed by Gauss, Laplace and Poisson to handle inaccuracies in data were not available to him. In truth I am not certain how Galileo himself handled the issue of errors beyond the extreme care he exercised in his measurements, but care he did exercise and it is no less important today to adopt the same approach.

Uncertainty, accuracy and precision

All measurements are subject to uncertainty. We make the distinction, however, between accuracy and precision.

- Accuracy is a measure of how close the measured value is to 'reality'; that is, what it ought to be.
- Precision is a measure of the reproducibility of the measurement.

There are two principle sources of a lack of precision: first, the measuring equipment itself may not be capable of measuring to a very precise degree, because, for example, the scale is too coarse; and, second, the method may be fundamentally inaccurate with some degree of variability on the inaccuracy so that repeated measurements of the same thing are unlikely to yield the same answer. Precision, then, can also be defined as:

- A measure of our confidence that the uncertainty on the measurement has been reduced to a minimum.

Accuracy and precision are described in greater detail below.

Accuracy in measurement

Consider, for example, the simple measurement of length. Some device will be needed for the measurement and the simplest is a rule. Rules are not natural objects, however. They have to be made and marked according to some accepted standard. So commonplace is this activity that we never stop to think whether the tape measure used by the attendant in the DIY shop, for example, to measure the length of wood is accurate or not. We have every expectation that it is: if we were to measure the length of wood required for a shelf and the DIY attendant were to cut it to the same length, it should fit neatly into the place we have reserved for it. It would indeed be a strange thing if the two measurements of 24 cm were different. However, it could happen if one of the rules were not marked correctly. Then, no matter how much care was taken over the measurement, one value will always be different from the other. If one 24 cm should really have been, say, 23 cm, as measured against a standard rule, this is an error of *accuracy*.

Precision in measurement

A measurement of, say, the length of a piece of wood to a value 24.0 ± 0.1 cm is more precise than 24.0 ± 0.5 cm. The latter says in effect that its length could actually be anywhere between 23.5 cm and 24.5 cm, so if some-one else were to measure, say, 23.7 cm there could be no quibble about the difference between the two measurements. However, in the former the limits are much closer, being 23.9 cm to 24.1 cm so an independent measurement of 23.7 cm would have to be considered to be a different value.

The precision assigned to any particular measurement is often a matter of judgement. A rule marked in centimetres is likely to be further subdivided into millimetres and possibly also in half-millimetres. In such a situation we have to make a judgement about where within the divisions the measurement lies. Beyond a certain limit we are faced with a guess and that is the point at which we define the precision. A measurement of 24.0 cm is just such a judgement. It may be that the wood has a rough edge or marked by a line that is itself a millimetre thick, so it is not possible to be sure *exactly* what the value is. The judgement is that it might be 23.9 cm, but it could be 24.1, so the measurement is set to be 24.0 ± 0.1 cm.

Precision and the role of randomness in measurement

Identifying the causes of the lack of precision is something that can cause the inexperienced experimentalist considerable difficulty. The precision identified above is essentially an *instrumental* precision. However, there are circumstances under which errors on a measurement are much greater than the instrumental precision. If the measurements are subject to a *random* uncertainty, the degree of randomness can far exceed the instrumental precision. For example, a stop-watch may be able to measure down to tenths or even hundredths of a second, but the point at which the clock is stopped may vary from time to time, especially if human intervention is required.

This is really a question of *accuracy*. The clock returns a precise, but inaccurate time. Contrary to the case of a miscalibrated instrument, which always returns an inaccurate measurement, the inaccuracy is not systematic but variable. Sometimes the clock will be stopped early and sometimes late. *Randomness arises from the variability in the accuracy of the measurement.* The exception occurs when we measure a process that is inherently random, for example radioactivity. We can detect precisely when a decay occurs but it is a chance event and therefore unpredictable. The average behaviour is predictable using statistics, though. Radioactivity is described in greater detail in Chapters 4 and 6.

▶ Physics and the origin of randomness

The distinction between the role of accuracy and precision in randomness is not always understood. There is a widely held view that all measurements are subject to random uncertainty, but this is not so; all measurements are uncertain, but not all the errors are random. This is a point worth emphasising again and again. Lord Rutherford, father of atomic physics, is reputed to have said that if you need to use statistics you had better get a different experiment. Random errors arise from the particular experimental method; change the method and the need for statistical analysis can be eliminated.

So how has it come about that current orthodoxy has it that all errors are random? Statistics is very often regarded as a branch of mathematics rather than physics, yet many of the earliest contributors to the field, such as the three already mentioned, namely Gauss, Laplace and Poisson, were actually physicists. Many since have taken randomness to be an inherent physical property whereas a physicist should make no such assumptions. If the errors are random then let a consideration of the physics show it. Rutherford clearly felt that the need to resort to statistics indicated a poorly designed experiment. Unfortunately there are still many texts dealing with error analysis that propagate the myth that all measurements are subject to random errors.

It will be shown that *if* the errors are random, the accuracy can be improved by taking repeated numbers of measurements and averaging over them. Furthermore, the more measurements taken the more the average value is expected to approach the 'true' value, and the smaller the error – called the standard error – on the mean. This obviously has implications for the methodology adopted. If the errors are not random repetitive measurements will not only be a waste of time but lead to false notions about the precision achieved.

Example 3.1   Randomness versus resolution

Consider the piece of wood measured earlier at 24.0 ± 0.1 cm. How is the precision to be estimated? Do we measure the wood one hundred times and take the average? Emphatically, no! It should be clear from the preceding discussion that this is fundamentally an error of resolution, not of randomness. The accuracy is determined by the calibration of the rule and it is impossible to see where any randomness might enter the measurement. If the orthodoxy of random errors were to be accepted, namely that it applies to all measurements, then it would be necessary to accept that for some unknown reason either the rule cannot be set in the same place each time or that it is not possible to make the same reading each time. Worse! possibly even the wood or the rule are randomly changing their length. As before, if the errors are random let a consideration of the physics show it.

Intuitively, we know that one measurement should be enough. Our initial intuition when faced with two measurements that did not agree would be that the measurement hadn't been made properly. Repetitive measurements in this situation are not only unnecessary but also wrong. Of course, the measurement could be forced to conform to the orthodoxy by saying what we *think* the value might be without quoting the errors, and repeating this several times. Thus one time it might be 23.9 and another 24.1, and on yet another 23.95 cm. In this way a series of results could be built up from which the mean could then be calculated. This is not very honest, though. Applying statistical methods to an artificially generated 'random' distribution would enable the derivation of a very 'precise' value because the more measurements taken the more precise the expected result.

Fundamentally the errors are not random and there is no point in treating them as such. The errors arise from the resolution of the rule. That is to say, it is not possible to resolve the measured value to a greater precision because of the way the rule is marked. On a different rule with different subdivisions the judgement of the precision might well be different, say ±0.2 cm, but fundamentally the two measurements would be expected to agree.

What about the case where a succession of different people take the measurement? A greater variability in the range of results might be expected under these circumstances because different people will exercise different judgement about where to place the rule and possibly about the sub-division of the scales. However, it is still not really random and it does not alter the fact that the uncertainty arises fundamentally from the inability to judge beyond a certain resolution exactly where the wood lies on the rule. If different people make measurements of the same object outside the quoted precision, either the precision does not accurately reflect their ability to judge the situation or other causes must be suspected.

Precision: a question of judgement?

It seems quite unscientific to assert that in problems of resolution very often it is necessary to make a judgement about the uncertainty. It should be borne in mind that the precision, and ultimately the error on the measurement, represents the confidence that can be placed upon a measurement. It is perfectly acceptable, indeed proper, to use one's judgement. Using one's judgement is certainly more scientific than applying an inappropriate mathematical treatment, by, for example, treating non-random errors as random.

Right-mindedness in measurement

Forcing a set of measurements that are not actually randomly distributed about a mean value but are arbitrarily chosen within some limits to appear random is a form of dishonesty. Dishonesty is, of course, a strong word. It can be used to imply a deliberate deception, and sometimes this happens in science. In my experience it rarely happens in physics, but if it does then of course it is wrong. That is a question of ethics, however, whereas I am concerned with the question of the right approach to the measurement. I have called it right-mindedness rather than honesty in order to emphasise that it does not necessarily arise from a deliberate attempt to deceive. Rather, it is a failure to appreciate the true nature of the measurement and to apply the appropriate methodology.

Science is essentially a public activity, and physics is no different. Even at undergraduate level your reports and laboratory work are going to be scrutinised by others. As a professional physicist you will be expected to publish reports and papers in the open literature that will be scrutinised by your peers and superiors. If you have adopted a false methodology your work will lack credibility. You will lack credibility. Right-mindedness in your approach to measurement embodies a personal commitment to openness and honesty in your approach that is essential if your work is to have any value.

Random errors and statistics

A single measurement subject to random uncertainties is inherently difficult to deal with. The instrumental precision may well exceed the experimental precision by some way, so an appeal to the physics will probably not give an accurate idea of the extent of the uncertainty. In fact there is no way of knowing the extent of the imprecision. The only sensible way of dealing with truly random errors is to take lots of measurements and see how they are distributed. A series of random variations will be describable by the Gaussian distribution, the so-called bell-shaped curve as illustrated in Figure 3.2. This graphical illustration demonstrates one of the important features of the random error; the results are distributed about a central value. The expected or expectation value are alternative terms. We can take this value to be the 'correct' value; that is the value that would be measured in the absence of any errors.

Visual inspection of this curve would lead you to believe that you could reasonably identify the central value of the distribution, but clearly a large number of measurements are necessary. With just a few measurements this would be inherently difficult. The fact of randomness means that some measurements will lie below the expectation value and some

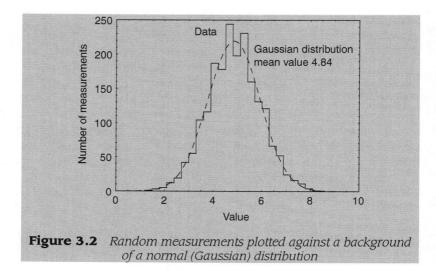

Figure 3.2 *Random measurements plotted against a background of a normal (Gaussian) distribution*

will lie above. With a lot of measurements there is a good chance that a good sample of both has been collected so by taking the average of all the results the effect of the fluctuations is cancelled out. With only a few measurements it will not be at all clear whether the results lie in the main above or below the expectation value, and the corresponding uncertainty on the mean must be much greater. The question then arises, 'How many measurements must be performed before this correct value can be estimated?'

A statistical analysis (see Chapter 4) will allow calculation of the mean value of the measurements (the central value), the probability of finding a measurement within a certain value of the mean, and the standard error on the mean, as the term is called. The standard error is a measurement of the precision with which the mean is known and increases with every additional measurement made. This is the essential feature of random errors; over enough measurements those that lie below the mean cancel out those that lie above. In order to know the quantity being measured to a specified precision we simply need to perform enough measurements.

In fact, very few measurements are truly random, and very few follow the Gaussian distribution exactly. One very important feature of the Gaussian distribution, however, is that it does not matter in practice whether the errors actually follow this distribution or not, it is correct to treat them as if they did. Thus even if the measurements are not truly random it is still possible to define a mean, a probability of finding a measurement within a certain value of the mean, and the standard error.

Propagation of errors

'Propagation of errors' is the term given to defining a precision on the outcome of the experiment. The outcome is often derived from the results of the individual measurements by some form of calculation. The individual measurements themselves are subject to some imprecision, and inevitably this propagates through the calculation to produce an imprecision on the outcome. You should be clear that it is the *precision* that is propagated through, not the accuracy. It does not matter whether the precision is derived from statistical considerations or considerations of resolution, provided it is realistic. Techniques for error propagation will be presented in Chapter 4.

Systematic errors affect the magnitude of the outcome but not the precision. There is no realistic way to treat systematic errors of accuracy other than to correct every inaccurate measurement and redo the calculation, but of course it is necessary to recognise first that a measurement may be in error. Such an error may well not be noticed, and often accounts for the fact that two independent measurements do not agree within the quoted precisions.

▶ Physics of measurement

So far the discussion has revolved around the types of measurement and the nature of errors. No consideration has been given to the physics of measurement except in passing. As the example of Archimedes' principle demonstrated, consideration of the physics can determine the choice of measurement. Even the simple measurement using a rule described above involves some physics. The physics here does not necessarily lie in the process itself, which in this case is no more complicated than reading a number off a scale and estimating the degree of precision, but in the *meaning* of the measurement. Measurement is all about meaning; are we measuring what we intend to measure?

For example, consider a metal rule. A metal typically has an expansion coefficient of the order of 10^{-5} K^{-1}. Taking this as our value for a given metal, this is the *fractional* change in length that the metal will undergo if the temperature is changed by 1 degree Kelvin. If the metal piece is precisely 1m long at say 23°C then at 24°C it will be 1.00001 m long. At 33°C it will be 1.00010 m long, and so on. A rule will not allow this sort of precision so normal temperature fluctuations can be discounted as a possible source of error in any experiment involving metal rules in the laboratory.

What about the case where it is necessary to measure something down to the micrometer scale? If the total length of the object to be measured is itself of the order of a few micro-metres then the demands of the measurement are very different from the case where the total length to be measured is of the order of metres but the measurement has to be accurate to the nearest micro-metre; that is, a precision of the order of 1 part in 10^6. In the first case, some sort of microscope will be needed to see the object, and the microscope can sensibly be calibrated to allow a measurement of an object under view. A measurement to an accuracy of a few per cent will probably suffice. In the second, a completely different technique will have to be employed, probably utilising optical techniques. Such measurements may well be accurate to the wavelength of light and will provide the necessary accuracy. Indeed, at one time the definition of the metre was based on optical measurements, and the history of the definition of the metre provides a very good example of the problems of measurement.

Example 3.2 Definition of the standard unit of length, the metre

The origins of the metre go back to at least the eighteenth century, at which time there were two competing approaches to the definition; the length of a pendulum having a half-period of one second, or one ten-millionth of the length of the earth's meridian along a quadrant (one-fourth the circumference of the earth). In 1791 the French Academy of Sciences chose the latter definition because the force of gravity varies slightly over the earth's surface, which alters the period of the pendulum, depending on where it is measured. It is worth noting that these are entirely arbitrary definitions that are only useful if the relevant measurements can be made to the required accuracy.

Therefore the metre was defined as 10^{-7} of the length of the meridian through Paris from the pole to the equator. Surveyors set to work to measure the distance, and platinum-iridium bars were constructed with appropriate engravings to represent the metre. The first prototype, made in 1874, was short by 0.2 millimetres because researchers miscalculated the flattening of the earth due to its rotation. Hence, in 1889, the first General Conference on Weights and Measures redefined the metre to be the distance between engraved lines on a platinum-iridium bar that would rest in Sèvres, in France.

In 1887, Albert A. Michelson had discovered how to adapt his interferometer to measure distance to a fraction of the wavelength of light. His method was so precise that the redefinition of the metre was actually obsolete, but it took a further 71 years before the definition was altered. In 1960 the General Conference on Weights and Measures adopted a definition based upon a wavelength of krypton-86 radiation, but 1960 also coincided with the invention of the laser. For practical purposes, it was again possible using a laser to measure the metre much more accurately than the definition would allow. The metre was demoted from a primary standard from which other units are derived to being itself a derived unit.

In 1983, the 17th General Conference on Weights and Measures decreed the velocity of light c is to be exactly 299,792,458 ms^{-1} and the metre is defined to be the distance light travels in $1/c$ seconds. In the 1970s, methods were developed for comparing the frequency of a laser to the frequency of an atomic clock, thereby allowing a very accurate way to measure the speed of light via the product of the laser's frequency and wavelength. The velocity of light could be measured to 4 parts in 10^9 because lasers could be made to achieve a frequency stability exceeding one part in 10^{10}.

This definition of the metre was practically useless. The metre is measured using wavelength, but the accuracy of the wavelength measurement is limited by the accuracy of the frequency measurement. Frequency measurement has proved very difficult in the past, but recently it has become possible with much simpler equipment to measure frequency to approximately 1 part in 10^{14}. This allows for even greater accuracy in the definition of the metre, and even makes the definition practical.

The definition of the metre is, like many quantities, artificial. Whether it is the length of a pendulum, a fraction of the Earth's diameter, a platinum bar, the wavelength of light, or ultimately the velocity of light that is used to define the unit, there is still an arbitrary definition at the heart of it. Even so, a sure knowledge of physics is required to ensure that the practical measurement fits the definition, which did not happen with the first definition in 1874. New discoveries, new inventions and new technology all play their part in extending the accuracy of the practical measurement.

Further reading: *Physics Today*, March 2001, vol. 54, pp. 11, 37.

Knowledge of the physics of what it is we are trying to measure is crucial to our deliberations. This is the first of the critical thinking skills described in relation to the research process; that of gaining knowledge which will inform us in our design of an experiment. Of course, we cannot know everything. The modern physicist is impeded by the proliferation in scientific literature published each year, but at the same time is assisted by the advent of electronic databases of published papers. Nonetheless, we often miss things, particularly older work. We can only be conscientious in our efforts to acquire whatever existing background knowledge is relevant, and hope that we have covered the field. Moreover, given the proliferation in published information it is all the more important that we be critical in our reading. Such considerations may not altogether apply at an undergraduate level where the necessity to read published journals is mitigated by the limited nature of the experimental work. Nonetheless, a critical approach to reading and acquiring background knowledge is still necessary.

▶ Accuracy, precision and new physics

The importance of accurate and precise measurements has already been illustrated by the example of the definition of the metre. The first standard was 0.2 mm short because the flattening effect of the rotation of the earth was not considered. It is an interesting question as to how this error was noticed, and one to which I do not know the answer. It is easy to suppose, however, that there must have been a number of careful measurements that did not tally with theoretical predictions, or possibly conflicting experimental observations. When faced with such a circumstance, the question immediately arises as to the reliability of the measurement. How much confidence can you place in the outcome? Are you absolutely certain that you have measured what you set out to measure? Are you certain of the errors on the result?

If the answer to these is 'yes'; if the measurements are accurate and the precision has been estimated correctly, these measurements correspond to the 'facts' of the real world, as described in the previous chapter. If enough variations in the 'facts' have been recorded without satisfactory explanation it must follow that some other property is influencing the results. Thus a new 'problem' is identified and a new field of research initiated. This is how the neutrino was discovered: a discrepancy in the momenta of nuclear particles following a reaction led to the postulation that a third particle must exist which carries off the excess momentum. Eventually the neutrino was discovered experimentally.

▶ Randomness and the experimental technique

As described, randomness in measurement arises from a fundamental variability in the accuracy of the measurement. For example, suppose you tried to measure the stiffness of a spring, also known as the spring constant, by attaching a mass to the spring and letting it oscillate. Theory tells us that the frequency of the oscillation depends on the square root of the stiffness, so by measuring one the other can be determined. If your measurement system comprises a stopwatch and the words 'go' and 'stop' you should not be surprised to find that there are large errors in your measurement. In fact, you would find it very difficult to make a judgement about the precision of the measurement based on a single event.

You could of course employ techniques such as counting 20 or 30 oscillations before stopping the clock, so whatever lack of accuracy results from your judgement about when to stop, timing is spread over a large number of oscillations. In effect you would be taking a mean as your result. Nonetheless, you should not expect a single result such as this to be very precise. How do you account for your reaction times at both the start and the end? Are they different? What about your judgement on when to say 'stop'? How reproducible will it be? This is a situation where the variability on the measurements is likely to exceed by far the resolution of the clock, but this can be tested simply by taking more measurements. You would find in all likelihood a spread of results ranging over several seconds.

The variability is an inherent feature of this method. No matter how sharp your reflexes or how good your eyes you could not guarantee to reproduce the measurement time after time. The only way to treat this problem is to take a number of readings and take the average. As far as the statistics are concerned it is not really important what distribution function applies to the results, though it may well show some interesting trends. It is possible to proceed, as discussed before, as if the distribution function were Gaussian and it will not make any difference to the answers about the average and the standard error on the mean.

Randomness and critical thinking

So far no critical thinking has been applied to this experiment. Following Rutherford we should be critical not only about the nature of random errors but about the method itself. What alternative method is there that can be used to eliminate the randomness? Particularly, the wisdom of employing a technique where the errors are so large, and essentially beyond our control can be questioned. If the orthodoxy about all measurements being subject to random errors had simply been accepted, then in all likelihood such questions would not have been asked: errors are expected, they exist, so get on with it. There is a way of dealing with random errors so if an alternative cannot be found then we are not ultimately lost. However, the necessity to perform lots of experiments in order to come to a sensible answer remains. Is it necessary? Is there no way to be more efficient? In critical thinking every aspect of the experiment can be questioned. It is possible, even, to ask whether systematic errors are introduced into the experiment.

Randomness and bias in measurement

Accepting that randomness follows from a variable lack of *accuracy* on an individual measurement, it is somewhat idealistic to hope that the variability is evenly spread about an ideal value. Variability may well exist, as the previous example illustrates, but the variability is a direct consequence of the experimental technique and may well come with an in-built bias. For example, it is easy to imagine that starting the clock and stopping it are different processes. There may be better warning for one than the other, for example if the start is timed to coincide with the release of the mass. An attempt could be made to eliminate difficulties such as this by waiting for the motion to reach a certain point before the clock starts and finishes so that as near as can be told the two are subject to similar random variability in reaction time. However, that does not take into account that we might try to predict the moment in order to counter the delay introduced by our reactions, and quite possibly we would predict the stopping differently from the starting.

It might well be that no such bias exists, but the point is that there is no guarantee. Furthermore, it is very difficult to discern from the outset whether any such bias exists or not. We are having to make a judgement of a moving object. We are not automata, but sentient beings with who-knows-what tendencies to systematize our behaviour. Whilst the statistics will account for the variability, they will not account for anything we do which tends to shift the measurement to either a longer or a shorter value. We are thus faced with an awkward situation; an experimental technique which may be biased, but without any reasonable means of estimating the bias. If it is impossible to recognise a bias beforehand, what about afterwards, when the measurement has been made and the result derived? Would it be possible to recognise any such error?

Unless there was an independent means of measuring the stiffness or unless the spring was intended for use in some application, such as an instrument for example, which was subsequently found to be in error the bias might never be recognised. Worse! such is the nature of the measurement technique it would be difficult to correct for a lack of accuracy other than to adjust individual measurements found to be in error. There is no sensible way to estimate the bias. This is quite unlike the case of the piece of wood and the DIY attendant where a lack of accuracy is easily recognised simply by comparing the two measurements and just as easily corrected by recalibrating the rules against a known standard. How can you calibrate your judgement? These are fundamental difficulties, and the reason no doubt that Rutherford cautioned against an experiment which required statistics.

Improved accuracy from multiple measurements

The fundamental difficulty with this experiment is that only one frequency has been measured. This value of frequency may be an average of many measurements but as we have seen, this does not guarantee to give an accurate value. The answer to improving the accuracy of the final result, which is not the frequency but the spring constant derived from the frequency, is to take several *independent* measurements. What constitutes an independent measurement in these circumstances? Possibly somebody else could perform the experiment, but this is probably not realistic if several people are required. Much better, the conditions under which the measurement was conducted can be changed. Chapter 2 dealt at length with functions, variables, and experimental conditions, and this provides the answer here.

The theory of a mass on a spring predicts that the frequency f of oscillation is given by:

$$f = \frac{1}{2\pi}\sqrt{\frac{k}{m}} \tag{3.1}$$

where m is the mass attached to the spring, and k is the spring constant. If the mass were to be changed and the whole process repeated a number of times, f could be plotted against $1/\sqrt{m}$ and the stiffness k derived from the slope, $(1/2\pi)\sqrt{k}$. Not all measurements of frequency would be equally accurate; the quicker the mass moves the harder it will be to judge, but this would be expected to manifest itself in the precision.

Plotting a graph in this manner has the advantage that the variation in the frequency with mass can be seen. It will be immediately apparent whether the data follows the straight line predicted, and it will be immediately apparent if there is some systematic bias that results in a constant offset. If there were a bias in the experiment it would be reasonable to expect it to act in all the measurements in a similar way and quite possibly a plot of f against $1/\sqrt{m}$ might not go through zero. A bias in a single measurement could not be accounted for, but a systematic bias appearing in several can be eliminated simply by taking the slope.

▶ Eliminating randomness

The lack of precision, and possibly accuracy, in this experiment is a direct consequence of the measurement method. If the variation of spring position with time could be measured in a different way without the variability there would be no need for the statistical methods employed in the previous example. Such a method might be, for example, to use a high-speed video with hundreds of frames per oscillation. It might be that the extreme values of the oscillation cannot be located precisely in time because the mass will slow as it stops and turns round so that there appears to be very little change from one frame to the next. However, this is now a problem in resolution. We will be able to identify at least one frame where the mass has not reached the extreme and another after the extreme has been reached (marked as $-a$ and $+a$, and also as $-b$ and $+b$ in Figure 3.3). If the frame rate is known the time at which the extremum occurred lies within the times corresponding to these two frames.

Quite possibly additional mathematical techniques could be employed to locate it further by, for example, direct differentiation of the position with respect to time. Calculating the differential close to the extrema may well prove difficult but it is only necessary to locate where the gradient goes to zero. Therefore if the gradient on either side of the extremum is known, interpolating between these points would provide a reasonably good estimate. If this technique was felt to be deficient a curve could be fitted to the data points as described in the last chapter and the function differentiated mathematically.

It is apparent that this technique would be much more precise than the last, and also more accurate provided the frame rate is calibrated correctly. It could still be advantageous to change the mass and re-measure

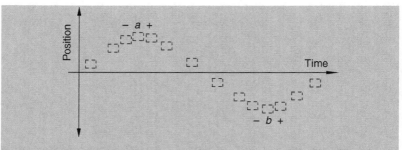

Figure 3.3 *Schematic illustration of a photographic record of the position of an oscillator as a function of time*

the frequency so that a graph could be plotted. Any imprecision in the data not accounted for would be immediately apparent in the scatter about the straight line, and the slope would give a more precise value of spring constant than a single measurement.

▶ Critical thinking and experimental design

The preceding pages have described various aspects of experimental design. Designing an experiment is but one part of the research process described in Chapter 1, but there are several critical skills involved, from gaining knowledge through to applying it. It should be clear by now that a critical approach to methodology can reveal much about the nature of the errors and the likely steps that we will have to take to ensure an accurate and precise outcome. Moreover, whether the outcome is accurate or not will not always be apparent. Unlike the undergraduate case where there is often a prior 'accepted' value, against which I warned in Chapter 1, it is necessary to rely entirely on the quality of the experiment for confidence in the result. As the outcome often cannot be checked against another measurement, the design of the experiment, the associated errors and their analysis, and the evaluation of the data are crucial. In short, the research process and the associated critical thinking are indispensable.

The essential elements in the design of an experiment can be summarised as follows:

- Knowledge of the physics and mathematical relationships defines the conditions.
- Several measurement options will probably be apparent, and some criterion is required to select one.
- Accuracy, precision, and ease of experimentation are important factors.
- Accuracy is the term used to describe the closeness of the measurement to the 'real' or expected value.
- Precision is a measure of the reproducibility of the measurement, and describes the confidence we can place on the measurement lying within specified limits.
- Randomness is not an inherent physical property but a consequence of the measurement technique.
- Randomness occurs when the spread of the results is greater than the resolution; that is, when the accuracy of a single measurement is low and variable.
- Random errors can be eliminated by redesigning the experiment.

- Random errors require repetitive measurements for high precision.
- Statistical methods are used to deal with randomness.
- Randomness does not have to follow the Gaussian distribution to be dealt with statistically.
- Errors of resolution can be determined from the instrument and can be applied to a single measurement.
- Critical thinking should help reveal a better experimental method.

▶ Instrumentation

The one important feature that has not so far been mentioned is instrumentation. It is not possible, nor sensible, to even attempt to describe laboratory instruments in any detail; there are just far too many. However, most instrumentation is now based on electronic measurements of one sort or another. Sensors will convert some effect into a voltage or a current which is what is ultimately measured. This discussion will therefore concentrate on the measurement of these properties and in particular the principles of input and output impedance, which are the most important parameters for most routine measurements.

The most important thing we want to know is whether we are measuring what we want it to measure. It is one thing to connect up an instrument, such as a voltmeter, and record a signal, but it is quite another to assume that the measurement is correct. There are circumstances under which an ammeter or a voltmeter will not work correctly, and it is as well to be aware of them. This is an error of accuracy, of course, but assuming that the instrument to be working correctly and accurately there is also the question of precision. It is very tempting with electronic instrumentation to assume that just because the measurement is displayed digitally to however many significant figures this is a sensible representation of the precision. Analogue meters don't suffer from this problem. Not only do we have to decide where the pointer stops but we also have to take care about the angle at which we view the meter, so thay seem inherently less accurate. Moreover, there is clearly a limit to the resolution set by the division of the scale that is not important in a digital instrument. The electronic instrument may indeed be more secure from some of these errors but the impression of increased accuracy and precision are often illusions.

Accuracy in instrumentation

Accuracy is concerned essentially with calibration. All commercial electronic instrumentation is calibrated at source but calibrations do drift over time. I am not suggesting that you need to learn how to calibrate an instrument; as an undergraduate you will not have that responsibility and as a professional physicist you would probably return the instrument to the manufacturer for the job. You need to be aware, though, that simply because an instrument displays a particular number this is not necessarily the value you should record. Unless you have specific reason to suspect that an instrument is giving false readings you would not normally take action, but there can be no harm in testing the instrument periodically.

The problem is essentially the same as that discussed in the last section: how would you know if the reading is false? Perhaps you will not know, and perhaps it will not be critical to your work. If it is critical, however, then you need to check the precision periodically. You could do this by, for example, generating a trial signal of known magnitude and checking the response. If it is an ammeter you are using measure the current flowing through a known resistor in response to a known voltage. However you might do it you could also take advantage of the fact that calibration errors are systematic in nature. If you can define an expected value and compare it with a measured value then you automatically have a method for correcting your readings, whether the instrument is recalibrated or not.

Precision and instrumentation

Precision is concerned with the number of decimal places for the measurement. Sometimes the manufacturer's specifications will tell you the degree of precision, but in other cases you will have to use your own judgement. With a digital instrument an estimate of the error can be difficult. More than ever, you need to use your judgement. Remember, the precision is a measure of the confidence you have on the measurement so be guided by what is reasonable and sensible. The following may be helpful:

- In a digital multimeter the display is often a fixed number of digits, with the decimal point moving according to the scale. The maximum precision is therefore fixed at 1 in 10^3 or 10^4.
- The actual precision of a measurement may be less than the maximum if the measurement is noisy. One particular decimal place may vary rapidly so that the measurement might not settle on a well defined value, and the uncertainty can encompass the whole of this particular decimal place.
- Computer-automated measurements will yield a very specific value, but this will not necessarily reflect the precision. The value returned will be the particular value at the time the data value was 'grabbed', but take the reading again and you may well find a different value. Precision can be improved by taking a number of measurements and averaging, with an estimate of the precision from the standard error (see Chapter 4).

Circuit loading in measurement

Even if the instrument has been calibrated there is still a possibility that the measurement will not be correct if the instrument *interferes* with the measurement circuit. This is always a possibility with electronics. There is no such thing as a totally passive instrument. It has to be connected to the circuit in order to record the data and in so doing will become part of the circuit. Under normal conditions of operation its effect on the circuit is negligible, but there are occasions when the opposite is true. The instrument is then said to *load* the circuit. The mechanism is the input impedance – effectively the resistance due to the instrument seen by the circuit – that all instruments possess and which appears in series with the instrument. If the input impedance is similar to the total impedance in the circuit, the total current flowing in the circuit or the voltage dropped across a component will both be affected. The measurement will therefore be false.

▶ Voltage measurement

An ideal voltmeter should have an infinite resistance. Ohm's law tells us that the voltage dropped across the resistor is given by the product of the current and the resistance, as shown in Figure 3.4.

For two resistors in parallel the total *conductance* ($1/R$) is found by adding the conductances. The voltage across each resistor is identical (it is physically impossible to have different voltages across the parallel resistors) so the current flowing in each resistor is different to ensure that an equality between the products iR is established. Hence:

$$\frac{1}{R} = \frac{1}{R_1} + \frac{1}{R_2}, \quad R = \frac{R_1 R_2}{R_1 + R_2} \tag{3.2}$$

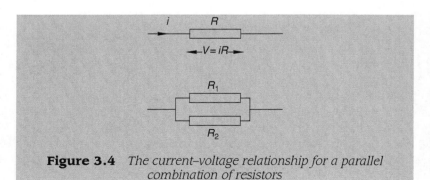

Figure 3.4 *The current–voltage relationship for a parallel combination of resistors*

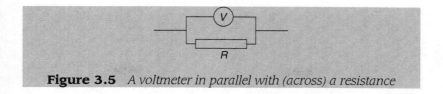

Figure 3.5 *A voltmeter in parallel with (across) a resistance*

If the two resistances are equal, then the total resistance of the combination is half the resistance of each individually. In other words, the total voltage dropped across the combination will be halved. If one of the resistances is much greater than the other, say $R_1 \gg R_2$ then $R_1 + R_2 \sim R_1$ so that $R \sim R_2$. In other words, the combination acts as if it consisted only of the lowest value resistor. In practice you need a minimum ratio R_1/R_2 of about 100 for this to be effective. In such circumstances the error introduced by the approximation is ~ 1 per cent.

This can be applied to the voltmeter by replacing one of the resistors by the instrument (Figure 3.5). Provided the circuit 'sees' a resistance due to the voltmeter which is much greater than the resistance being measured there is no problem. Voltmeters do not have infinite resistance, however. Typically, for digital instruments the resistance, known technically as the input impedance, is $1\text{M}\Omega$. This means that if an ordinary voltmeter is connected across a resistor of more than $10\text{k}\Omega$ significant amounts of current will start to flow through the voltmeter, and thereby reduce the voltage across the resistor. This is called *loading the circuit*. At $10\text{k}\Omega$ the loading is 1 per cent and at a load resistance of $100\text{k}\Omega$ 10 per cent of the current flowing in the circuit would be passing through the voltmeter. For a load resistance of $1\text{M}\Omega$ this value would rise to 50 per cent. In short, the voltage measured would be half of what would be expected.

Voltage measurement across high impedance: the buffer amplifier

So, how do you measure a voltage across a high impedance load? Obviously we need to connect something across it that has an even higher impedance. The buffer amplifier (Figure 3.6) has a very high input impedance and a low output impedance – the effective series resistance of the output – but a gain of 1, known as unity gain. It doesn't amplify the voltage but the low output impedance can be connected to an ordinary voltmeter. It acts as a buffer between the high impedance load and the instrument, hence the name, and is therefore ideal for measuring high impedances that would normally be loaded by a $1\text{M}\Omega$ input impedance. With input impedances in the range

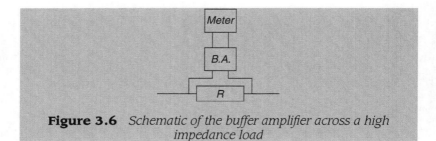

Figure 3.6 *Schematic of the buffer amplifier across a high impedance load*

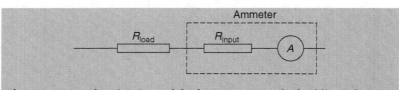

Figure 3.7 *The circuit model of an ammeter (dashed line) showing the effective series resistance*

$10^{10}\Omega$ and $10^{15}\Omega$ depending upon the particular amplifier used, the voltage can be measured across very high impedances without loading the circuit. Buffer amplifiers are normally constructed from operational amplifiers (commonly called 'Op-Amps'), which are integrated circuits supplied as discrete devices.

▶ Current measurement

For accurate current measurement, the opposite argument applies. Ammeters are connected in series with a load, so any input impedance loads the circuit. It is important therefore that the input impedance of the ammeter is at least 100 times smaller than resistance in the circuit or else the ammeter contributes to the total resistance and affects the current flowing. Ammeters are not just used in circuits containing relatively high resistances but are general instruments that may also be connected across low impedance loads. The input impedance of the ammeter should therefore be as low as possible.

Measuring very low currents

Just as a voltmeter has its limitations, an ammeter is also limited. For very low currents flowing in a circuit, of the order of a nano-amp or less, measurement of the current is tricky. An ordinary ammeter will not be sensitive enough, and passing the current through a resistor to measure the voltage developed across it requires high impedances to develop a measurable voltage. The operational amplifier has to be used again, but this time the gain does not need to be unity, as above. It can be set higher, to values of 10, 100 or even 1,000, say, in order to amplify the signal and to enhance resolution. Commercial instruments called electrometers work on this principle and can measure currents as low as 10^{-15}A. This is phenomenal; it is no more than about 10,000 electrons per second and is close to the theoretical limit.

▶ Frequency-dependent impedances

Similar considerations apply to frequency-dependent measurements. In particular, the impedance of a capacitor, X_C, varies inversely as the frequency, so at low frequencies the impedance can be very high, especially if the capacitance is small. Time-dependent signals such as sinusoidal or square-wave voltages are usually recorded on an oscilloscope which typically has an input impedance of 1MΩ. Again, a buffer amplifier must be used for high-impedance measurements, but with the additional consideration that the frequency response of the amplifier must not only allow the signal to pass but must not interfere with the phase of the signal, if the phase information is important. A knowledge of the electronics is needed to ensure that the amplifier will work at the intended frequency.

Coaxial cables

Coaxial cables are commonly used to connect to oscilloscopes. A coaxial cable consists of an inner conductor (usually a single wire) sheathed in plastic which is in turn sheathed by a flexible metal braid. The braid is usually connected to earth in any measurement so the inner conductor, which usually carries the signal, is shielded from external electromagnetic sources of noise. The design of the cable means that there are two additional factors that need to be considered in any measurement; capacitance and impedance. The geometry of a dielectric sheath separating two conductors should be recognisable as a capacitor, and coaxial cables typically have capacitances ranging from 10–50 pF/ft. In many measurements the capacitance of the cable will not interfere with the circuit, but if the cable forms part of a capacitance measurement the total capacitance of the cable will add to the measured capacitance and represents an offset.

The impedance arises because the coaxial cable is a transmission line. An electromagnetic signal of frequency 300MHz will have a wavelength of 1 metre, which is similar to the length of a cable in a circuit. Therefore signals of this sort of frequency would be radiated into free space if conducted along open wires, which act as efficient aerials when the length is similar to the wavelength. A transmission line is formed by having two parallel conductors and can transmit a wave down the length of the conductors efficiently and without loss of power to radiation. All transmission lines have a characteristic impedance Z, which in the case of the coaxial cable at high frequency, is given by:

$$Z = \sqrt{\frac{L}{C}} \tag{3.3}$$

Impedances of coaxial cables range from 25Ω to 120Ω, and possibly higher, but common values are 50Ω and 75Ω. Oscilloscopes have an input socket with a characteristic impedance of 50Ω to match that of the cable.

Impedance matching

It can be shown mathematically that for any wave the impedance can be defined in terms of the propagation velocity of the wave. If the wave is incident on a boundary where the velocity changes, part of the wave will be reflected and part will be transmitted. The purpose of impedance matching, therefore, is to ensure that none of the power is reflected back down the cable.

Impedance matching is not just about maximising the power transfer, however. Any impedance mismatch, caused, for example, by connecting the inner conductor to a bare wire, causes a reflection which will propagate back down the wire and then reflect back at the first impedance mismatch encountered. Some small time later, probably no more than a few nanoseconds depending on how long the cable is, this reflection will appear on the oscilloscope screen. Figure 3.8 illustrates the mechanism. A signal (1) incident on a mismatch ($M2$) is partially reflected and partially transmitted (2). At mismatch $M1$ this is partially reflected (3) which in turn is partially

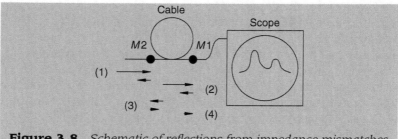

Figure 3.8 *Schematic of reflections from impedance mismatches*

reflected off $M2$ (4). Signals (2) and (4) therefore appear on the oscilloscope screen separated by the round-trip time between $M1$ and $M2$. The round-trip time will normally be a few nano-seconds, so if the duration of the signals is similar the reflections off mismatches can represent a serious distortion of the signal.

Time constants

Quite apart from reflections due to impedance mismatches, the time-constant of the circuit is one of the principal limitations in high-speed measurement. The time-constant τ is the product of the effective series resistance R and capacitance C, both of which are found by combining the impedances of the individual circuit elements to produce the equivalent series circuit consisting only of these two components. Ideally τ should be a factor of 10 smaller than the shortest event to be measured, otherwise the signal may well not rise to its true value.

▶ Bandwidth

In electronics the bandwidth is defined as the frequency range over which the gain falls by 3dB, which is equal to half the maximum value. In other words, the gain is equivalent to the full width at half-maximum (FWHM). For high frequencies the gain falls to zero. Amplification can still occur outside the specified bandwidth, therefore, but the amplification will not match the specifications. For example, it is possible to record signals at a frequency of 300MHz on an oscilloscope with a bandwidth of 100MHz, but the voltage displayed will probably be less than the true voltage in the circuit. Operation outside the bandwidth must be treated with care, but at the very least the time-dependent behaviour can be observed even if the amplitude may be awry.

▶ Output impedance

Just as measuring instruments have an input impedance that can affect the circuit, voltage and current sources have an output impedance that can also affect the circuit. As with the input impedance, the output impedance appears in series with the instrument.

- For a voltage source the output impedance should be as low as possible. The voltage source may be connected across a low value of resistance and has to deliver effectively an infinite current. No voltage source can, of course, deliver such a current, but the principle is that the resistance of the voltage source should not limit the current. Hence the resistance should be as low as possible.

- For a current source the output impedance should be very high. In effect the current is determined by the total resistance of the circuit and this should be due to the current source. Therefore the output impedance should be higher than any other component in the circuit.

▶ Summary

This chapter has described the considerations involved in designing an experiment. In particular, it has been shown that a number of different possible measurements can be undertaken according to your objectives. A detailed understanding of the physics is crucial not only to the choice of measurement but also to an understanding of the sources of errors. Archimedes' principle has been used as a specific example but the ideas apply generally.

The mathematical treatment of errors has been left to Chapter 4. Here the emphasis has been on understanding the difference between accuracy and resolution. Errors of accuracy can be:

- Systematic, counteracted by calibration.
- Variable, counteracted by the use of statistics if the resolution exceeds the accuracy.

It is emphasised that the physics of the problem should dictate whether statistical methods are to be used to treat data.

The principles of electronic instrumentation have also been described, as the majority of laboratory equipment is increasingly based on electronic sensing. The key principle is to ensure that the instrument does not interfere with the circuit. Ideally:

- A voltage source has a low impedance and delivers a high current.
- A voltage measurement has a high impedance and draws no current.
- A current source has a high impedance and maintains a voltage.
- A current measurement has a low impedance and drops no voltage.

4 Statistics, Physics and Probability

Statistics means probability, and at first sight probability and physics seem to have little need for each other. Some areas of physics, such as classical mechanics and thermodynamics, seem to have a solidity and certainty about them that lends the appearance of factuality. Classical physics is solidly deterministic. Cause leads to effect and the effect is precise and well known. The motion of bodies subjected to forces, the mechanical advantage of pulleys and levers, the work done by an expanding gas, and so on, are all definite and calculable effects upon which the technology of the industrial revolution was based. There appears to be no room for probability here.

Although the theories may be deterministic, it does not follow that real effects can be measured to the same precision as theory predicts. The origin of modern statistics in physics can be traced to the eighteenth century, and the need to handle imprecise measurements in astronomical observation. This led to the development of the method of least squares, which has already been mentioned in Chapter 2 as a criterion for computational determination of unknown parameters. It is relatively easy to perform large numbers of calculations with digital computers today, but in the time of Legendre and Gauss, working in the late eighteenth and early nineteenth centuries, no such resource was available. These physicists had to rely on their own mathematical ability to develop algorithms capable of solution by hand for determining unknown quantities in experimental observation. This they did, and laid the framework for much of the use of statistics in experimental data analysis.

The majority of this chapter will be devoted to the statistics of measurement, but it is worth noting that many ideas in physics incorporate statistical concepts, a development initiated during the nineteenth century. The kinetic theory of gases and statistical mechanics are two such, and this chapter will start with this aspect of statistical physics. This is not the way the subject developed historically; analysis of errors came first, followed by developments in the mathematics, which then found application in physics. It is the intention here to reverse this process; to use familiar ideas in physics to introduce the statistics so that the use of statistics to analyse experimental data can be understood.

▶ Probability: a historical perspective

The statistical analysis of astronomical observation developed apace in the nineteenth century. Key to this development was the notion of a distribution of errors, building on the work done earlier by De Moivre and other mathematicians who developed the notions of chance and probability. It was De Moivre who first described, in 1733, what is now called by physicists the Gaussian distribution, but in statistics is often referred to as the normal distribution. Much work was done by mathematicians of the early eighteenth century and before, such as Laplace and Euler, on what De Moivre called the 'Doctrine of Chances'.

A cursory examination of their work on statistics suggests a preoccupation with gambling, which of course is not physics, but the concepts were later to be applied to physical problems. The principle of, say, defining the chances that four aces will be drawn from a pack of 52 cards can be applied directly to the determination of the way in which impurity atoms can be arranged in a solid. The statistical mechanics of Boltzmann and Gibbs, sometimes referred to as statistical thermodynamics, was developed in the last half of the nineteenth century, and Boltzmann was awarded his doctorate in 1866 for the development of the kinetic theory of gases.

It was early in the twentieth century that the notion of the statistical analysis of data really took hold, with the work of Sir Ronald Aymer Fisher. Indeed, in 1922 Fisher gave a new definition of statistics. Its purpose, he said, was the reduction of data, by which he meant the process of reducing an amount of information so large it cannot easily be comprehended, to just a few meaningful numbers which represent the distribution of the data and from which the essential properties of the data can be derived. Fisher contributed to the development of data analysis using statistics, and invented methods to determine whether parameters estimated from different data sets are statistically different. Undoubtedly, much of the present understanding of statistical analysis stems from Fisher's time.

It was also clear by this time that statistical analysis was firmly part of mathematics and divorced from its background in physics. Fisher worked for many years in the field of agricultural research, where very large samples of seemingly randomly distributed data are generated simply by the fact of natural biological variability; data, for example, on leaf sizes, crop yields, or bacterial growth and populations. These sorts of data are inherently different from the data of interest to physicists.

The difference is in the nature of the randomness. Biological variations are probably an inherent property of the system, but, as described in Chapter 3, randomness is a physical consequence of the measurement technique; imperfect observations arise from imperfect measurements. *The purpose of the analysis is to see whether, by virtue of repeated measurements, we are better able to arrive at a more accurate value than a single measurement alone will allow.* Statisticians such as Fisher was, on the other hand, are more interested to know whether natural phenomena are appropriately described by a particular mathematical function and then to perform a detailed statistical analysis in which data reduction occurs. Data reduction in physics is intended to calculate the average value of a set of imprecise measurements to arrive at an estimate of the precision, which is a much simpler task.

▶ Probability in physics

Probability in physics is invoked when there is an inherent unpredictability. This is certainly the case in the quantum view of the world, but very often it arises from the sheer scale of the problem, and there are many examples of processes that are fundamentally statistical in nature, but where we deal with macroscopic averages. For example, the laws of thermodynamics, together with the gas laws of Boyle and others, are macroscopic manifestations of microscopic phenomena; that is, the small-scale random motions of the atoms. The entire kinetic theory of gases is in fact a statistical construct; the number of electrons in a solid available for electrical conduction is determined by probability; radioactivity and nuclear decay provide yet another example in which we have no knowledge of individual atoms (that is, when precisely will it decay?), but given the large numbers of atoms within a radioactive source it is possible to use statistics to calculate average decay rates and other useful properties, such as the half-life.

Consider Fick's first law of diffusion by way of a detailed example.[1] Fick's law appears in many areas of physics. It is used to represent, among other things, the flow of electrons in a semiconductor or metal, the motion of atoms in a gas, and the flow of heat, the transport of impurity atoms through

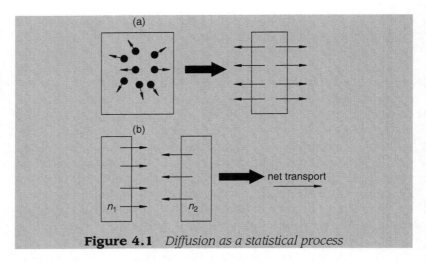

Figure 4.1 *Diffusion as a statistical process*

a solid lattice. The basic form of the equation expresses the flux, or flow per unit time per unit area, as being proportional to the concentration gradient, that is:

$$J \propto \frac{dn}{dx} \tag{4.1}$$

where J represents the flux and n is the concentration. This can be understood very easily with reference to random motion of particles, as shown in Figure 4.1. Consider a small volume containing particles with random velocities as shown in (a). This can be represented as a small element of volume with an equal flux of particles in both directions. As we are only concerned with motion in one dimension – the horizontal – it doesn't matter that some of the particles have a vertical component to the velocity. If this element is next to another with a different number of particles, as in (b), there will be an exchange of particles between the elements according to the difference in concentrations of particles. There will therefore be a net transport in the direction shown.

The total rate of transport depends on the average velocity of the particles, which of course depends on how easily they move through space. This is all contained in the diffusion coefficient D, the constant of proportionality which is found experimentally. Finally, the transport occurs in the direction of the lower concentration, that is, a negative gradient, so

$$J = -D\frac{dn}{dx} \tag{4.2}$$

Diffusion is therefore a statistical process. It is impossible to know which atoms are moving in which direction, so a microscopic description is not possible. There is no explicit reference to probability in the above, but it is there nonetheless. A different example makes the connection between statistical phenomena and probability more clearly.

The velocity distribution in gases

As already mentioned, it is impossible to know the velocities and positions of each and every molecule in even a small volume of gas. A single molecule is a different matter altogether. Classical mechanics is deterministic. Given the forces acting on a particle it is possible to trace out the trajectory exactly so that at any particular time the position and velocity can be calculated, taking into account collisions with the walls of the container. These can be dealt with using the principle of conservation of momentum and kinetic energy. The latter is not a general law but applies in the case of molecular collisions.

If another particle is present this might also collide with the first particle, and in order to be able to treat the problem it is necessary also to know this particle's trajectory so that the moment of collision can be predicted. Moreover, it is desirable to establish some connection between the particles so that an approach leading to a collision is an integral part of the solution. Such a connection may be established, for example, by formulating the problem in terms of the *separation* of the two particles, rather than a position relative to some external reference. In this way the equation of motion for each particle is linked to the equation of motion for the other. Such equations are said to be coupled because each has a solution dependent on the solution of the other.

If a third particle were to be introduced the complexity of the problem would increase as its motion would have to be coupled to the motion of the other two. As particles are added there comes a point where analytical solutions are just not possible. The problem has to be solved numerically. That is to say, the motion of each particle is computed piecewise in small time intervals. Even this becomes very difficult, if not impossible, with a relatively small number of particles, perhaps a few hundred. The limiting factor is the speed and size of available computers. Today, calculations with such small numbers of particles, usually arranged in a solid rather than a gas, are routinely performed. This field of physics is generally known as molecular dynamics, and can be very informative, but it is not possible realistically to model more than a few hundred particles. Even though computational technology will vastly improve there will likely be some limitation. In 1 cm^3 of gas there are approximately 10^{20} atoms or molecules. There is no way that this number of particles can be modelled on a computer. It would be virtually impossible to set up the calculation, never mind solve the problem.

It is necessary, therefore, to resort to statistics and probability. We do not *know* the properties of individual particles, but by assigning to each a probable position x and speed v, the average properties of the ensemble can be calculated. The speed is of interest as this depends only on the magnitude of the velocity and not its direction, which has already been described as random. There will be as many negative velocities as there are positive so the average velocity will be zero but the average speed will not.

▶ Probabilities and distributions

The probability for the position of the particle is easiest to deal with. There is no physical reason to suppose that the particles will be distributed anything but evenly, unless some external factor such as a sound wave propagating through the gas were to cause the molecules to move in an ordered manner. Any concentration gradient will lead to diffusion, as described, so fluctuations in particle density will tend to even themselves out. This means that the probability of finding a particle in any position is identical to the probability of finding a particle in any other position. The particles are said to be distributed uniformly.

The distribution of speeds is a little more difficult. Maxwell and Boltzmann[2] showed that the number N_v of molecules, out of a total number N, having a speed between v and $v + \Delta v$ is given by:

$$\frac{\Delta N_v}{\Delta v} = \frac{4N}{\sqrt{\pi} v_m^3} v^2 \exp\left[\frac{-v^2}{v_m^2}\right] \tag{4.3}$$

where v_m is the most probable speed. It can be shown that at any temperature T,

$$v_m = \sqrt{\frac{2kT}{m}} \tag{4.4}$$

Equation (4.3) is a little daunting. It may not be immediately apparent how probability enters into this equation, but this distribution is an entirely statistical concept. Figure 4.2 shows the speed distribution calculated for two different temperatures using the parameters shown. The mass corresponds roughly to that of hydrogen. The most probable speed corresponds to the speed at the maximum of the distribution. What is immediately noticeable is that the range of speeds increases with temperature, and that the maximum value of the distribution decreases as the temperature increases. This is because the total number of molecules is constant and if

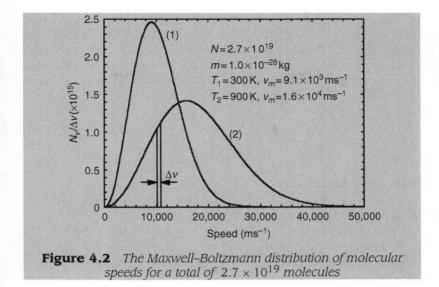

Figure 4.2 *The Maxwell–Boltzmann distribution of molecular speeds for a total of* 2.7×10^{19} *molecules*

there is a greater probability of finding a molecule with a higher speed there must be a smaller probability of finding a molecule with a lower speed.

What this distribution means in practice is the following. If we could take a container of the gas and somehow separate out the molecules according to their speed, we would find 1.1×10^{15} molecules with a speed lying between v and $v + \Delta v$, as shown in Figure 4.2. The range Δv represents a 'velocity window' and if we could move that window along from $v = 0$ to $v = 4,500$ (for curve 2) we would see the number of molecules within the window increase up to ~1.4×10^{15} at the maximum velocity before dropping down to zero again at the higher temperature. We don't know which molecules will have these speeds, we only know that a certain number given by equation 4.3 will have them.

Probability as frequency

Equation 4.4 for the speed distribution is actually a number distribution. It gives the number of molecules within a speed range. We convert this to probability by dividing through by the total number of molecules,

$$\frac{(\Delta N_v / \Delta v)}{N} = \frac{4v^2}{v_m^3 \sqrt{\pi}} \exp\left[-\frac{v^2}{v_m^2}\right] \tag{4.5}$$

This is one of the most commonly accepted definitions for probability; it is the *frequency of events expressed as a fraction of the total*. Put another way, if instead of sorting the molecules according to their speed we were to pick them out one at a time and examine their speed, the number of times we would record a speed lying between v and $v + \Delta v$ would be, as before, 1.1×10^{15}. The difference is that this would be a random process. We could not pick out just those molecules with the speed we require unless we had some prior knowledge as to which they were, so it is a matter of chance. We define that chance according to the fraction of molecules that have the required speed, that is, $1.1 \times 10^{15}/2.7 \times 10^{19} = 4.1 \times 10^{-5}$. On average, just over 4 out of every 100,000 molecules will have the speed we are looking for.

Probability distributions

A probability distribution, as the term implies, is the distribution of probabilities within a system. It is not the same as the distribution of events. In counting molecules, as above, we would not find that after every 25,000 molecules one of them has the speed we require. The sequence of events is usually random. Consider one of the simplest systems possible, that of a tossed coin where there are only two possible outcomes; heads or tails. There are only these two probabilities in the system, and this therefore represents the probability distribution. However, toss a coin repeatedly and almost certainly it will not alternate between heads and tails, for although that arrangement would satisfy the distribution of probabilities it would in fact represent an ordered system; it would imply a correlation with the preceding event so that the coin would have to 'know' somehow that a tail had been thrown for the next event to be a head.

A truly random process, where there is no bias towards one result or the other, would exhibit sequences of heads or tails, but on *average* we would expect equal numbers of each. This raises the immediate difficulty of the size of the sample from which real probabilities are measured. Suppose, for example, five heads were thrown in a row. Based on these five events alone you would conclude that there is no chance of throwing tails. In a larger sample the long sequences are less important. In a sequence of, say, a thousand throws, it would be possible to throw 1,000 heads consecutively, but not very likely. We would expect 500 of each on the basis of the probability distribution, so sequences of heads should be reasonably well balanced by sequences of tails. For example, 495 heads might be thrown to 505 tails.

Probability and certainty

One of the properties of a probability distribution is that it represents all possible outcomes. In the definition of probability as frequency probability is a fraction and the sum of all the fractions must be unity. That is to say, the sum of all the probabilities in the distribution add up to certainty, because any event, whatever it is, will correspond to one of the outcomes contained within the distribution. Mathematically, if the probability that an event i will occur is p_i, then

$$\sum_i p_i = 1 \qquad (4.6)$$

▶ Statistics in measurement

The discussion of probability distributions and the frequency of events naturally leads on to the problem of measurement, the statistics of errors, and how we can analyse the measurements to be sure of our results. We need to be clear about what we are measuring. In the case of a tossed coin it is a probability, but in other cases, for example the astronomical measurements upon which modern statistics were founded, it will be the average value of several numbers. The essential problem for the experimentalist is then: *given an experimentally determined outcome from statistical analysis, what physical meaning can be attached to it?*

In the case of a coin it seems straightforward. We know the ideal probability is 0.5 and from the previous example the *experimental* probability is 0.505 for throwing a tails. We have taken a large sample of a thousand throws so clearly there is a slight bias in favour of throwing a tails. Or is there? It is necessary to estimate a precision on this probability. It will be shown that for this example the precision is just over 3 per cent based only on the number of measurements. That is to say, the experimentally determined precision could be anywhere between 489 and 521 even for a perfectly fair coin, so a value of 505 is perfectly acceptable.

This is the essential principle of all statistical analysis. In Chapter 3 randomness was described as originating from a variable accuracy so that one measurement is different from another. We can have no idea about the variability from just one measurement so we take many. We expect that the more measurements we take the better our ideas about the variability so that we can express the result of the measurement with some confidence. In statistical terms these are defined as the *location* and the *scale*, and to calculate them, we need to know the mathematical form of the probability distribution that governs the measurements. This is called the *limiting distribution*, because it is the distribution we expect will be achieved in the limit of a very large sample. Different limiting distributions have different location and scale parameters, and limiting distributions of interest to physicists will be described in this chapter.

Limiting distributions and expectation values

The principal limiting distributions used in experimental physics are the Poisson distribution, used for counting random events such as radioactive decay, and the Gaussian distribution, sometimes called the normal distribution. In both cases the location parameter of interest is the mean but they represent physically different things. It will be shown that the mean of the Poisson distribution is related to the probability of the event being counted whereas the mean of the Gaussian distribution represents a central value about which a range of measurements is distributed. The physics of what we are trying to measure must inform the statistics used.

Practical location and scale parameters

The mean is one location parameter, the median is another. This defines the point at which half the values in the distribution lie below and half lie above. Very often the mean and the median coincide, but not always. Which location parameter therefore constitutes the most useful? It might be argued with some justification that if half the measured data lie below a certain value and half lie above, this value constitutes an effective outcome of the statistical analysis. However, it also seems intuitively sensible to take the mean, and indeed this is the most commonly accepted location parameter.

Similarly, there is more than one possible scale parameter, but in practice we use the variance and the standard deviation. It should be noted that these scale parameters are not necessarily the same as the precision, but are closely related to it as will become clear. Rather, scale parameters represent the scale of the distribution. If the result of the ith measurement is x_i, and the mean is $\bar{x}$, then the variance is:

$$s^2 = \sum_i^N \frac{(x_i - \bar{x})^2}{N} \tag{4.7}$$

and the standard deviation is:

$$s = \sqrt{s^2} = \sqrt{\sum_i^N \frac{(x_i - \bar{x})^2}{N}} \tag{4.8}$$

where N is the total number of measurements. You may be familiar with some of these concepts, and in particular the use of the symbol σ and σ^2, instead of s and s^2. Here the distinction is made that s is used for quantities derived from measured data whilst σ is reserved for the

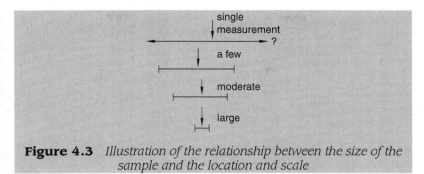

Figure 4.3 *Illustration of the relationship between the size of the sample and the location and scale*

analogous quantities appropriate to the limiting distributions, but in many texts no such distinction is made.

It is important to remember that these are not the only scale parameters, but they are the most commonly used and the most useful. They are related inversely to the number of measurements so the more data we take the smaller the scale parameter, and hence the more confidence we can have in the outcome of the measurement. It is important to note that the location parameter (the mean) might not change very much with an increase in the number of measurements (Figure 4.3). This can be summarised as follows:

- Clearly a single measurement means very little; we don't know what the precision may be.
- Two or three measurements allow a mean to be calculated but the error is large.
- A moderate sample may allow a sensible average but still the error may be large.
- A large sample may not change the average value much but the error will decrease.
- We would expect the average value to be an accurate and precise representation of the measured data in the limit of a large sample.

▶ The purpose of statistics in data analysis

The meaning of statistical analysis is therefore clear, its purpose is to provide a means of estimating:

- a sensible value of the location parameter when the measurements are subject to apparently random variability;
- the scale parameter, or the error, or precision, on this quantity.

In the case of a coin we can evaluate the location parameter we expect to find – a probability of 0.5. In most measurements, however, there is no prior expectation value, as the location parameter of the limiting distribution is called, so we have to assume that our experimental result relates somehow to the ideal distribution. Therefore, *the experimental location parameter is an approximation to the expectation value and the error is a measure of the reliability of this approximation.*

Statistics and decision-making

Calculation of a location parameter allows conclusions to be drawn which tell us something about the physical world. In the example of the coin, 1,000 measurements have been performed and a probability of throwing a 'tails' of 0.505 ± 0.016 has been calculated. That is, we expect the 'true' value to lie between 0.489 and 0.521. The decision to be made in this case is, 'Does this accord with our assumption about the nature of the process under investigation?', The answer is clearly, 'Yes, it does'. Had we made 1,000 measurements and recorded a probability of, say, 0.460 to a precision of 3 per cent, the expectation value would lie outside the range of the measurements, and we would have to conclude that there is some external factor influencing the outcome of a throw so that on average more heads than tails are thrown. In other words, the coins are biased.

In experimental physics the decisions made on the basis of the statistical analysis will be similar, but there may be more than one decision to be made. We may decide, for example, whether an experimentally measured quantity agrees with:

- a theoretical prediction, or
- a separate, independent experimental measurement.

The principle is the same, however. The error derived from the analysis is a measure of the *confidence* we place upon the measurement, and if a second value of the same quantity lies outside our confidence limits we must conclude that the two are *statistically significantly different.*

Example 4.1

Suppose we have two independent measurements of some property, say $p_1 \pm \Delta p_1$ and $p_2 \pm \Delta p_2$. If $(p_2 - p_1) > \Delta p_1$ (Figure 4.4) then p_2 is a statistically different measurement from p_1. However, p_2 also has a precision, and suppose that it encompasses p_1 so by this criterion they don't seem statistically different. What might we conclude?

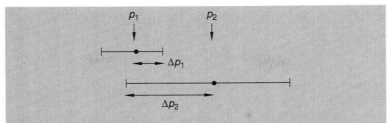

Figure 4.4 *Measured probabilities and precisions where p_2 seems statistically different from p_1, but the precision on p_2 is so large that it encompasses p_1*

- Δp_2 is overestimated so that p_2 is more accurate than we think (p_2 and p_1 are different).
- Δp_1 is underestimated so that p_1 is less accurate than we think (p_2 and p_1 are not different).
- Both Δp_1 and Δp_2 are wrong (we don't know).
- Either, or both, p_1 and p_2 are unrepresentative, which can happen if the sample is too small (we need more data).
- All four parameters represent the data, in which case the lack of precision in p_2 would lead us to give this much less weight compared with p_1.

In order to decide which of these is correct it will be necessary to review the measurements and the manner in which they have been conducted to see whether there are any obvious flaws in the method. If not, there is no alternative to repeating the measurement, possibly increasing the sample size at the same time in order to test the effect on the estimated parameters.

▶ Probability distributions of interest in experimental physics

The normal distribution is described first, as it applies most generally to the statistical treatment of errors. However, it is difficult to justify the normal distribution in a simple way, and it appears to be nothing more than a formula with no easily discernable physical origin. Some faith that it can be justified is required. On the other hand a simpler distribution, such as the binomial distribution, can be comprehended through a physical example.

The binomial distribution itself is of limited interest to the physicist but it has the particular property that under certain circumstances it approximates to the Poisson distribution, which finds application in nuclear physics. These three distributions will be described.

The normal, or Gaussian, distribution

Physicists usually refer to this distribution as Gaussian, but statisticians tend to refer to it as the normal distribution. The two terms are interchangeable. Variability in experimental measurement is assumed to be governed by the normal distribution, which has the functional form:

$$P_{x,\sigma} = \frac{1}{\sigma\sqrt{2\pi}}e^{-1/2\frac{[x-\bar{x}]^2}{\sigma}} \tag{4.9}$$

Here, $\bar{x}$ is the mean and σ is the standard deviation. For the normal distribution the standard deviation is a natural measure of the width of the distribution.

The normal distribution as a continuous distribution

The normal distribution is a *continuous* distribution function. The variable x can take any value, and this can cause problems in the analysis of experimental data. Consider leaf sizes, to borrow an example from agricultural research. We could take a very large number of leaves from the same plant species and measure one particular dimension of the leaf. Now pick one leaf at random. We want to know the probability that this leaf will have a certain dimension, but it may well be that there is no other leaf *exactly* the same. This leaf seems unique, and because it's picked at random probably every other leaf is unique. Probability doesn't seem to apply. However, our leaf will be very similar to other leaves, so instead we ask what the probability is that we will find a similar leaf. If leaf sizes range from, say, 45 mm to 55 mm, this can be divided into 0.5 mm intervals. A measurement of 49.6 mm will therefore fall within the interval 49.5 to 50.0 mm. The choice of intervals is somewhat arbitrary, but if the intervals are too small the numbers within them may fluctuate widely, and if they are too large the shape of the distribution will be lost. Such a plot is called a histogram, and was illustrated in Figure 3.2.

Experimental measurements and the normal distribution

Rarely in physics is it necessary to map out distributions in the manner just described. Of course, if we want to know whether a particular set of data is described by the normal distribution, then there is no alternative to painstakingly sorting the data and plotting the histogram. For most purposes, however, variations in data can be viewed as the result of some inaccuracy in the measurement technique, as already described in Chapter 3. It is not actually very important whether the data is truly described by the normal distribution or not. An experimental mean and standard deviation (equations (4.7) and (4.8), can be calculated no matter what the form of the distribution.

Scale and cumulative probability distribution functions

The scale function describes the width of the distribution. Its importance lies in the fact that it allows definition of the probability of a single measurement lying within the specified range. This is often called the *confidence limit*. For the Gaussian distribution the confidence limit is 68 per cent for one standard deviation, as determined from the cumulative distribution function (cdf), which is found by integrating over the entire limiting distribution. Integration from the lowest range to any point on the distribution maps out the cdf, as shown in Figure 4.5. The cdf (dotted line) is shown for a normal distribution (solid line) with a mean of 3 and a standard deviation of $\sigma = 0.7$.

Mapping the standard deviation (double-headed arrow) onto the cdf allows evaluation of the probability of *not* finding a measurement within the scale parameter. The probability of finding a single measurement within the scale is, from the cdf, $(0.84 - 0.16) = 0.68$, that is, 68 per cent. The probability of not finding the measurement within the scale is

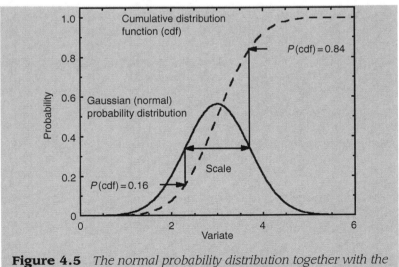

Figure 4.5 *The normal probability distribution together with the cumulative distribution function*

0.32 or 32 per cent. Of course, this has been evaluated for a limiting distribution, but what of experimental data? We *assume* that an experimentally measured distribution is a sample of a particular parent distribution, so we apply the properties of the limiting function to the distribution of data. In other words, we assume that s and σ are identical.

▶ Data rejection

How do we use the confidence limit in practice? Suppose you have a set of measurements of the same quantity collected from various sources. The measurements are independent of each other so it is possible to calculate the mean and standard deviation of the set. Now you take a single measurement on your apparatus of the same phenomenon. Does your measurement lie within one standard deviation of the mean or not? If not, there is reason to doubt the measurement, but remember, there is still a 32 per cent chance that the measurement will fall outside the scale parameter. The measurement cannot be discarded but it should be treated with caution.

What about a set of data you have collected in the laboratory. There is no substitute for care in experimental technique, but occasionally mistakes are made. The question then arises, should the data be rejected or not. If it is obvious that the data is erroneous then the answer is undoubtedly, 'yes'. However, the difficulty with statistically distributed data is that even valid data covers a wide range. The cumulative probability function can be used to show that the probability of a single measurement lying within 2σ of the mean is 95.4 per cent and 99.7 per cent within 3σ. Therefore, if the normal distribution is assumed to describe the data, anything lying outside 3σ can be taken to be improbable and therefore erroneous.

Standard error on the mean
The standard deviation does not provide an estimate of the error on the mean. This is given by the standard error, expressed as:

$$\frac{\sigma}{\sqrt{n}} \tag{4.10}$$

where n is the total number of measurements. This is derived from the central-limit theorem, which says that even if data itself is not normally distributed, the means from different samples will be normally distributed. For example, take a data set consisting of 1,000 different measurements. The mean and standard deviation can be calculated and because the data set is large we can be confident that the mean is an accurate reflection of the data. Call this the master set. However, suppose we take the first fifty of these numbers as also comprising a data subset. The mean and standard deviation for these can also be calculated, and because the subset is small there might be a slight difference from the mean and standard deviation of the master set. However, we would not expect the difference to be large. We could make 20 similar subsets from the master set and each would behave in a similar way, with the means and standard deviations varying slightly. However, the mean of all the means will be identical to the mean of the master set. We might suppose therefore that the mean of the master set is a better reflection of the data than the mean of any one subset. In other words, the mean of the master set is more precisely known, but we need to put a figure on this precision.

It is not simply the standard deviation of the master set, because the standard deviations of all the subsets will be similar to the standard deviation of the master set; after all, they come from the same parent distribution. However, if the standard deviation of the distribution of the *means* of the subsets is taken, it turns out that this is simply the standard deviation of the master set divided by the square root of the total number of measurements (Equation (4.10)). This means in practice it is not necessary to subdivide a set of data and calculate the mean and standard deviation of the means; the whole data set can be treated as a master set from which the mean, standard deviation, and standard error can be calculated quite simply.

▶ The binomial distribution

If instead of a wide-ranging variability the outcome of an event is dichotomous, that is, it can be one of two alternatives, the binomial distribution is used. The toss of a coin is a simple example. In a single throw the coin will land either heads or tails, but the distribution can also be applied to the case where multiple throws are considered. Thus we want to know the probability that r outcomes (heads) will occur in n events (throws), irrespective of the order in which they are achieved. This is given by a function equivalent to the rth term of the binomial expansion (see Appendix 3), hence the name. Assume again that the coin is equally likely to land heads-up as it is to land heads-down. Counting the number of heads as the outcome, in one throw there are two possible outcomes; in two throws, three possible outcomes; in three throws, four possible outcomes, and so on, as illustrated in Table 4.1.

These events are not equally probable, however. The probability of throwing three heads out of four throws is less than the probability

Table 4.1 *Possible outcomes for a tossed coin*

Throws	Possible number of heads returned
1	0, 1
2	0, 1, 2
3	0, 1, 2, 3
4	0, 1, 2, 3, 4

Table 4.2 *The number of ways of achieving an outcome*

Outcome	Sequence (H = heads up, T = tails up)	Ways
0 heads	TTTT	1
1 head	HTTT, THTT, TTHT, TTTH	4
2 heads	HHTT, HTTH, HTHT, TTHH, THHT, THTH	6
3 heads	HHHT, HTHH, HHTH, THHH	4
4 heads	HHHH	1

of throwing two heads out of four throws. The order in which the events occur is not important in deciding the outcome, so if the first two throws give heads whilst the last two give tails, this is the same as if the middle two gave heads whilst the first and the last gave tails. There are still only two heads. Clearly, though, there are more ways in which two heads can be achieved than there are ways in which three heads or four heads can be achieved. The ways in which the outcomes can be achieved are called configurations. The configurations are illustrated in Table 4.2 for four throws and five outcomes.

Out of 16 possible ways of landing, there are six ways in which two heads can appear. Using the definition of probability as frequency, the probability of this outcome is 6/16, or 0.375. For 1 head or 3 heads, antisymmetric outcomes, the probabilities are equal at 1/4, or 0.25. For the extremes, that is, 0 heads or 4 heads, there is only one possible way of achieving either result, and the corresponding probability is 1/16 = 0.0625. Figure 4.6 illustrates schematically the probability distribution, where the bars represent the respective probabilities that r, the outcome, will be 0 heads, 1 head, 2 heads, and so on up to 4 heads, from a single throw of the four coins. Again, this distribution is not necessarily equal to the distribution of actual

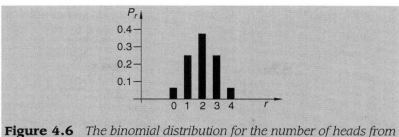

Figure 4.6 *The binomial distribution for the number of heads from four coins*

events in a coin tossing experiment; it is the *limiting* distribution derived from purely logical considerations.

Mathematically, the binomial distribution is expressed by:

$$P_r = \frac{n!}{(n-r)!r!}p^r(1-p)^{(n-r)} \tag{4.11}$$

where p is the probability of an individual event. In this case $p = 0.5$ because the coin must land either heads or tails. Hence the probability of the coin landing tails up is $1 - p = 0.5$.

Equation (4.11) can be understood as the product of a the number of configurations and a probability of occupancy. The configurations exist as a matter of geometry, and can be regarded as states, one of which must be filled on throwing the coins. This is the approach taken in much of physics. In the Maxwell–Boltzmann speed distribution (equation (4.3)), for example, the number of ways that a particular speed can be obtained is a geometrical problem and is referred to as a state, whilst the exponential is related to the probability of one of these states being occupied. The occupancy probability for the binomial distribution is represented by the power term in $p^r(1-p)^{n-r}$, as shown in Table 4.3. The probability of sequential events is simply the product of the probabilities of each event, and Table 4.3 lists the probabilities for the first three outcomes in Table 4.2 to show that, irrespective of the order of events, the probability has the same form $p^r(1-p)^{(n-r)}$.

If $p = 0.5$, this expression is equal to $1/16$ for $n = 4$ irrespective of the value of r, and represents the denominator in the calculation of the probabilities from Table 4.2. The number of configurations is derived by considering the combinations of r identical objects among n different ways. Consider first the case where the heads of the coins can be distinguished; that is, the heads are labelled H_1 and H_2 etc. If there are n coins, the first head can be distributed n ways. There are now $(n-1)$ coins among which to distribute the second head, that is, it can be distributed in $(n-1)$ ways. Similarly

Table 4.3 *Detailed explanation of the possible ways a number of heads can be thrown from four coins*

Outcome r	Sequence	Probability	Function
0 heads	TTTT	$(1-p)\,(1-p)(1-p)\,(1-p)$	$p^0(1-p)^4$
1 head	HTTT	$p(1-p)\,(1-p)\,(1-p)$	$P^1(1-p)^3$
	THTT	$(1-p)\cdot p\cdot(1-p)\,(1-p)$	
	TTHT	$(1-p)\cdot(1-p)\cdot p\cdot(1-p)$	
	TTTH	$(1-p)\,(1-p)\,(1-p)\cdot p$	
2 heads	HHTT	$p\cdot p\cdot(1-p)\cdot(1-p)$	$p^2(1-p)^2$
	HTTH	$p\,(1-p)\,(1-p)\,p$	
	HTHT	$p\cdot(1-p)\cdot p\cdot(1-p)$	
	TTHH	$(1-p)\cdot(1-p)\,p\cdot p$	
	THHT	$(1-p)\cdot p\cdot p\cdot(1-p)$	
	THTH	$(1-p)p\cdot(1-p)\cdot p$	

a third head can be distributed in $(n - 2)$ ways, and so on. The number of ways W in which r heads can be distributed among n coins is therefore:

$$W = n\cdot(n-1)\cdot(n-2)\ldots(n-r+1) \tag{4.12}$$

which can be simplified further as,

$$W = \frac{n!}{(n-r)!} \tag{4.13}$$

For two heads ($r = 2$) thrown from four coins ($n = 4$), for example, this expression incorporates the possibilities H_1H_2TT or H_2H_1TT, and so on, and gives 12 possible ways. If there were three distinguishable heads from four coins there would be 24 ways in which the coins could land. Each of the possible sequences in Table 4.3 can be similarly expanded by the number of permutations among the r heads, which is in fact equal to $r!$. It follows therefore that if *no* distinction is made between the heads the number of ways of distributing the heads is reduced by a factor equal to the number of permutations. Hence:

$$W = \frac{n!}{r!(n-r)!} \tag{4.14}$$

The mean of the binomial distribution

The mean, or average, value of any series of numbers $r_1 \ldots r_N$ is given by:

$$\bar{r} = \frac{r_1 + r_2 + r_3 + \ldots + r_N}{N} \tag{4.15}$$

where the bar denotes the mean. For the binomial distribution, however, the values $r_1 \dots r_N$ are limited in range. If the total number of times we record the number of heads, N, is large, say > 100, so that the experimental distribution has a reasonable chance of approximating to the limiting distribution, a repetition of values is inevitable. Taking the example of the number of heads thrown from four coins, r is in fact limited to values of 0, 1, 2, 3 and 4. Taking the general argument of m coins,

$$\bar{r} = \frac{n(0)r(0) + n(1)r(1) + n(2)r(2) + \dots + n(m)r(m)}{N} \tag{4.16}$$

which can be further simplified using the notion of probability as frequency, to become:

$$\bar{r} = p_0 r(0) + p_1 r(1) + p_2 r(2) + \dots + p_m r(m) \tag{4.17}$$

In short, the mean is the summation of a number of terms in probability, that is:

$$\bar{r} = \sum_{r=0}^{m} r \cdot P_r = \sum_{r=0}^{n} r \cdot P_r \tag{4.18}$$

where now we put $m = n$, to accord with previous expressions. It can be shown that the summation in (4.18) is equivalent to $n \cdot p$, which intuitively must be the case. Consider again the number of heads from four coins, where $n \cdot p = 2$. Suppose the set of four coins is thrown one hundred times in order to map out a distribution. In total there will be four hundred coins thrown, and the expectation value would be a total of 200 heads. Spread over the 100 attempts, these 200 heads represent an average of 2 heads per attempt, as given above.

Skewed distributions

The preceding example gives a symmetrical distribution about the mean. This is only the case for $p = 0.5$. If, for example, the coins were not fair so that the chances of throwing a head was, say 0.25, the average of the distribution would be $n \cdot p = 1$. Inevitably, there would be a greater number of outcomes of lower value. The probability of throwing 0 heads would be 0.316, and that of 1 head would be 0.422, 2 heads would have a probability of 0.211, 3 heads a probability of 0.047, and 4 heads a very small probability of 0.003. If, on the other hand, the probability of throwing a head were, say, 0.75, the distribution would be skewed in the other direction, with 3 heads being the most probable, and also the mean, outcome. The greater the imbalance between p and $(1-p)$, the greater the *skewness* of the distribution (see Figure 4.7).

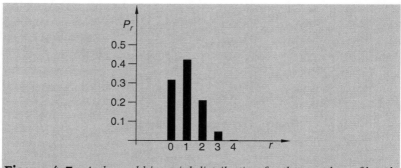

Figure 4.7 *A skewed binomial distribution for the number of heads from four coins with p = 0.25*

Applications of the binomial distribution

The binomial distribution can be used in any counting experiment where the number to be counted is subject to some randomness. An example is in nuclear experiments, where a certain number of counts may be recorded in a given time interval. In another time interval a different number of counts may be recorded. The mean count rate can be determined from this data. However, nuclear counting is best described by the Poisson distribution, which is a special case of the binomial distribution, and will be left to a later discussion.

One of the most common applications of the binomial distribution is in quality control. This is not strictly physics, but as a physicist you might, for example, find yourself employed in a semiconductor fabrication laboratory with responsibility for estimating fabrication yields; that is, the number of devices on a chip that work compared with those that don't. This is counting statistically distributed numbers and therefore an application of the binomial distribution. The dichotomy is 'device that works' or 'device that doesn't work', and the same criterion can be applied to any manufacturing control. Is the product good enough or not?

It is comforting to know that our intuition is backed up by mathematical analysis. We would expect intuitively that if we were to sample devices at random, the total proportion of devices that fail will give us a measure of the yield. This is what the mean of the binomial distribution also tells us. If r is the number of failures, and n the total number of devices tested, then from the definition of probability as frequency $p = r/n$ is the probability of finding a device that doesn't work. However, from the binomial distribution np is the mean of the distribution, that is, the average number of failures, and this is simply equal to r, the total number of failures recorded. It follows that $1 - p$ is the device yield. However, the question arises, what constitutes a random sample? It would be possible to test every one hundredth device, for example, but this would not necessarily pick up a systematic manufacturing failure in which every nth device were faulty.

The binomial distribution allows for an alternative method of collecting the data, that of batch testing. Take a batch consisting of n devices and evaluate the number of failures r within the batch. If M batches are tested, there will be $r_1, r_2, r_3 \ldots r_M$ recorded failures from which the distribution can be constructed. The mean can also be evaluated using the probabilities given by $p(r) = f(r)/M$, where $f(r)$ corresponds to the frequency of r. Hence the probability of failure can be calculated simply by dividing the mean number of failures by the number of devices tested in a batch.

The variance of the binomial distribution

It is possible now to define the scale parameter for the binomial distribution. The variance may be calculated according to equation (4.7) that is:

$$\sigma^2 = \sum_{r-0}^{n} (r - np)^2 P_r \tag{4.19}$$

because the mean is simply np. It can be shown that,

$$\sigma^2 = np(1 - p) \tag{4.20}$$

For the example given earlier of 505 heads from 1,000 throws, the variance is 250 and the standard deviation is $\sim$15.8, that is just over 3 per cent. In other words, the mean is 505 ± 16, which easily incorporates the expectation value.

▶ The Poisson distribution

The Poisson distribution is a limiting case of the binomial distribution and is important for counting phenomena, especially in nuclear physics. In the binomial distribution we want to know the probability of finding r events from n tries, but now we want to know the probability of counting r events in a specified time from n atoms. This is, in fact, more general than it seems. Yariv[3] has used Poisson statistics to consider the number of photons emitted by an ensemble of atoms in order to calculate the noise in a laser; that is the rate at which photons are emitted that are not part of the laser mode but nonetheless appear in the laser output. However, he also used the general treatment to derive the blackbody radiation spectrum – the wavelength distribution of electromagnetic radiation emitted by a perfect hot body as a function of temperature – first derived by Planck. This is one of the cornerstones of modern physics which helped to usher in the quantum era, because Planck found it necessary to assume that the energy is emitted in discrete amounts in order to be able to calculate the spectrum. A photon is either emitted or it is not, which, of course, is the assumption underlying the use of the Poisson distribution and underlines once again the place of statistics in physics.

The principal difference between the two distributions is that r is small compared with n in the Poisson distribution, whereas the maximum value of r is equal to n in the binomial distribution. For example, we may have 10^{20} atoms, but if the count rate reaches even a few thousand per second, so that the total count may reach a few tens or even hundreds of thousands, this is still negligible compared with the number of atoms. Hence the term $(n - r + 1)$ approximates to n, as do all the terms preceding this in the series, and (4.12) becomes:

$$n(n - 1)(n - 2)\ldots(n - r + 1) = \frac{n!}{(n - r)!} \approx n^r \tag{4.21}$$

Hence the probability becomes:

$$P_r = \frac{n^r}{r!} p^r (1 - p)^{(n-r)} \tag{4.22}$$

where the approximation has been dropped as being of no significance if $n \gg r$. There are a number of steps and approximations involved in developing (4.22) further, but eventually we arrive at,

$$P_r = \frac{\beta^r e^{-\beta}}{r!} \tag{4.23}$$

Here $\beta = n \cdot p$ is the mean.

Properties of the Poisson distribution
The Poisson distribution is an approximation of the binomial distribution for small values of p. It is to be expected, therefore, that just as the binomial distribution is skewed, the Poisson distribution is also skewed. Figure 4.8 illustrates the distribution for four values of the mean: 0.5,1,2 and 4. As the mean increases, the distribution tends to symmetry. For a mean value as low as nine the distribution is only slightly asymmetrical and in fact can be approximated by the normal distribution.

Standard deviation of the Poisson distribution
The assumption that $p \ll 1$ has implications for the standard deviation. The term $(1 - p)$ in (4.20) can be set equal to unity so the standard deviation simply becomes equal to the square root of the mean, β

$$\sigma = \sqrt{\beta} \tag{4.24}$$

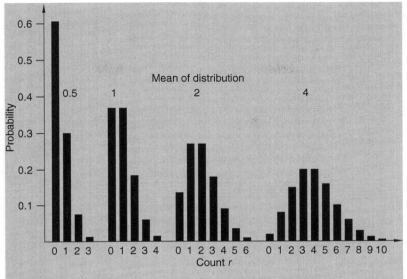

Figure 4.8 *Examples of the Poisson distribution for means of 0.5, 1, 2 and 4*

The mean of the Poisson distribution

The Poisson distribution is calculated on the basis that we measure the number of successes in n trials. Each atom in a radioactive sample constitutes a trial and each decay constitutes the success. The product $\beta = n \cdot p$ therefore represents the total number of atoms in the sample n multiplied by p, the probability that an atom will decay within the time of our measurement. This is just the total number of counts. *The total count in a Poisson distribution represents the best estimate of the mean of the distribution.* In order to understand this, we need to consider the Gaussian approximation to the Poisson distribution.

▶ Gaussian approximation to the Poisson distribution

Figure 4.9 shows the Poisson distribution (symbols) for three different means of 9, 10 and 15 overlayed by Gaussian distributions (lines) for means of 8.75, 9.75 and 14.75. If the Poisson mean exceeds 9, the Gaussian distribution is a good approximation, but the means are slightly different.

The standard deviations correspond almost exactly, however. What does this mean for our estimate of precision? The precision on the mean of the

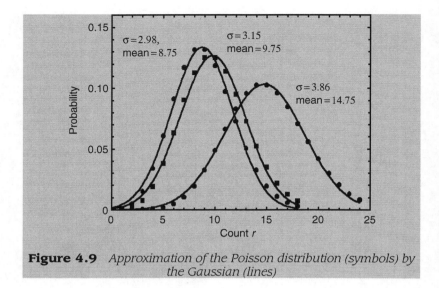

Figure 4.9 *Approximation of the Poisson distribution (symbols) by the Gaussian (lines)*

Gaussian distribution is given by the standard error, which is smaller than the standard deviation, so is the Gaussian more precise than the Poisson distribution? Intuitively the answer must be no. The data is the data and simply changing the statistical distribution should not alter the precision. The answer is that these are in fact different distributions. One way of mapping out the Poisson distribution is to record the number of particles incident on a detector in a given time and repeat the measurement m times. Each symbol in Figure 4.9 represents the number of times a particular count has been recorded. The mean of these counts, α, and the experimental standard deviation, s, correspond to the mean and standard deviation of a Gaussian distribution.

The total number of counts recorded is $\beta = m\alpha$ with an error $= \sqrt{\beta}$ which corresponds in fact to the standard error $\delta\alpha = s/\sqrt{m}$. Counting for a longer period of time is itself a form of averaging. Counting 10 times for one minute is in principle no different from counting once for 10 minutes. The total number of counts will be the same and the mean count rate will be the same. For example, if the average count rate was 10.0 counts per minute, as determined from 10 separate counting experiments, the standard deviation of the Poisson distribution $s = \sqrt{10} = 3.2$, but the standard error of the Gaussian distribution is $3.2/\sqrt{10} = 1$. However, in a single count lasting 10 minutes the total count would be 100, and the precision on this would be $s = \sqrt{100} = 10.0$. That is, if we were to count for another

10 minutes we would confidently expect to record a total count in the range 90 to 110, which corresponds to a count rate somewhere between 9.0 or 11.0 counts per minute, that is 10.0 ± 1.0. In short, nothing is to be gained from repetitive measurements using the Poisson distribution. Only the total count is important, and this count represents the best estimate of the mean.

▶ Error propagation

Having discussed the errors associated with different statistical distributions, it is now possible to consider how these errors are propagated. As discussed in Chapter 2, an error in x, say Δx, translates to an imprecision in y of Δy, according to equation (2.40). Where a function of two or more variables exists, each variable contributes via a partial differential, that is:

$$\delta y = \frac{\partial f(x,z)}{\partial x}\delta x + \frac{\partial f(x,z)}{\partial z}\delta z \tag{4.25}$$

The manner in which these terms are added depends upon the origin of the errors. Although the emphasis in this chapter has been on statistical errors rather than errors of resolution it is as well to distinguish between the two.

Propagation of errors of resolution: maximum error
The maximum error is found simply by adding error terms. As an example consider the error on an area A. The variables x and z represent the length and width, and the area is just the product of these. It is easy to show that $\partial A/\partial x = z$, and, vice versa, $\partial A/\partial z = x$. Therefore,

$$\delta A = z\delta x + x\delta z \tag{4.26}$$

This is the same result as you would get if you multiplied $(x + \delta x)$ by $(z + \delta z)$ and ignored the term $\delta x \delta z$, which is negligible in comparison with the other two.

Propagation of statistical errors
Where the outcome is a function of two variables, say x and z, both of which are statistically distributed, it follows that the function itself is statistically distributed. This distribution must have a variance, a standard deviation, and a standard error. If each pair of values x_i and z_i differ from their means by δx_i and δz_i then each value y_i will differ from its mean by:

$$\delta y_i = \frac{\partial f(x,z)}{\partial x}\delta x_i + \frac{\partial f(x,z)}{\partial z}\delta z_i \tag{4.27}$$

The experimentally determined variance can be found by the mean square of these deviations, that is:

$$s^2 = \frac{1}{n}\sum_{i=1}^{n}(\delta y_i)^2 = \frac{1}{n}\sum_{i=1}^{n}\left(\frac{\partial f(x,z)}{\partial x}\delta x_i + \frac{\partial f(x,z)}{\partial z}\delta z_i\right)^2 \tag{4.28}$$

The term in brackets can be expanded and cross products in $\delta x \delta z$ ignored. The summation leads to the variance in both x and z, leaving:

$$s^2 = \left(\frac{\partial f(x,z)}{\partial x}\right)^2 s_x^2 + \left(\frac{\partial f(x,z)}{\partial z}\right)^2 s_z^2 \tag{4.29}$$

If there are more variables an additional term for each variable is added.

Propagation of co-variant errors

The above analysis applies only to independent variables. Where a co-variance is suspected, that is, the value of x affects the value of z, the only sensible way to treat the errors is to calculate all the individual values of the final function and treat the distribution.

▶ Least squares

The method of least squares has proven to be very valuable in the past. It is still very important, but knowledge of the details is not so. Nonetheless, an appreciation of the principle is still useful.

Suppose a variable x exists which is statistically distributed; that is, we have a series of values x_i, $i = 1$ to n. Is there a value of x, say x_{min} such that the sum of all the terms $(x - x_{min})^2$ is minimised? There is such a value, and it turns out to be none other than the mean $(\bar{x})$. In other words, minimising the squares is equivalent to finding a value of the variable about which the deviations are normally distributed. This property turns out to be very useful for determining the best line through a set of points.

Least-squares curve fitting

This technique is also called linear regression. This does not imply that a straight line is to be fitted to the data, but that the function is linear in the unknowns. If the unknowns are A, B, C, etc, functions of the form $A \cdot B$, A^2, $\sin(B)$ and so on are non-linear in the unknowns and the method of least squares does not apply. The proliferation in scientific software in recent years means that you are unlikely ever to use the method for least-squares fitting of a straight line or polynomial; the software will do it all for you, but, again, it is useful to appreciate the principle, and to have the formulae just in case.

For example, suppose a function of the form,

$$y = A + Bx \qquad (4.30)$$

describes a physical reality. An experiment has been performed to measure y as a function of x, from which A and B are to be determined. The individual values of y are not normally distributed as their magnitude depends on the value of x. However, the function provides an ideal, or expectation value of $y(x)$ so for each measurement it is possible to define a deviation $y_i - y(x)$ for the ideal line, which corresponds to the best values of A and B. These deviations may be assumed to be distributed normally about zero, that is the sum of the squares is minimised. Under these circumstances, if

$$\Delta = N \left[\sum x_i^2 \right] - \left[\sum x_i \right]^2 \qquad (4.31)$$

where N is the number of points, then:

$$A = \frac{\left[\sum x_i^2 \right] \left[\sum y_i \right] - \left[\sum x_i \right] \left[\sum x_i y_i \right]}{\Delta} \qquad (4.32)$$

$$B = \frac{N \left[\sum x_i y_i \right] - \left[\sum x_i \right] \left[\sum y_i \right]}{\Delta} \qquad (4.33)$$

The uncertainty in these is given by:

$$s_A^2 = s_y^2 \frac{\left[\sum x_i^2 \right]}{\Delta} \quad s_B^2 = \frac{N s_y^2}{\Delta} \qquad (4.34)$$

where the experimental error on y, s_y, is given by:

$$s_y^2 = \frac{1}{N-2} \sum_{i=1}^{N} (y_i - A - Bx_i)^2 \qquad (4.35)$$

The formulae given above apply only to a straight line; if applied to a set of data they will yield the best straight line through the data. Similar formulae exist for a parabola and other polynomials, but it is easier to use software. It would also be possible to fit a polynomial through the same set of data, and the method of least squares will give the best values for the unknowns in that polynomial.

Which curve?

In Chapter 3, an example was given of data subject to some fluctuation and, similar to the above, the question was asked, 'Is a straight line more appropriate than a curve?' Equations (4.31) to (4.35) will tell you the best straight line through the data but will not tell you whether a straight line is the best function. So, on the basis of the data alone, is it possible to tell whether a curve better describes the spread of points than a straight line. To an experienced physicist the answer may seem obvious. One of two responses will be offered. First, what does the physics say? The great majority of experiments undertaken are done with some knowledge of the predictions of the physics as to the outcome. This must always be the guide. Second, if it is not immediately obvious that one curve is better than another there can be little justification for assuming the more complicated behaviour (curve) when the simplest behaviour (straight line) will describe the data. Is there any scientific justification for this view?

In order to answer this question it should really be rephrased: 'Which is the better description of the data *given* the assumption that the data is distributed normally about the expectation value?' The answer then is clear. The method of least squares can be used to fit both a straight line and a curve, and the distribution of the residuals, that is, the deviations, will provide the answer. A histogram of the residual frequency against magnitude should be normally distributed for the most appropriate line. However, a simpler method will often be instructive. That is, to plot the residuals against the independent variable. If the fitted curve is not the most appropriate some systematic variation in the residuals will be apparent. If the curve is the most appropriate the residuals will appear to be randomly scattered, as in Figure 4.10. You can satisfy yourself that if a straight line were to be fitted to this data the residuals, defined as $\delta y = (y_{expt} - y_{fit})$, would first be positive, then negative, and then positive again, which corresponds to a systematic variation.

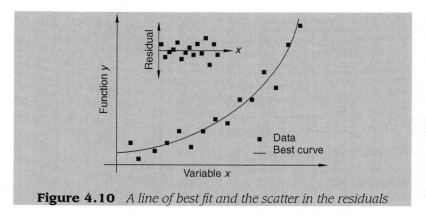

Figure 4.10 *A line of best fit and the scatter in the residuals*

▶ Summary

This chapter has looked in some detail at statistical concepts of use to the physicist. Of necessity, some concepts have been dealt with in sufficient depth to give a clear understanding whilst others have been brushed over and only the necessary details given. In particular, this chapter has shown that:

- Statistical concepts are relevant to the development of physical theories when large numbers are involved, for example for the kinetic theory of gases.
- The most common definition of probability is based on the concept of the frequency of an event.
- The sum of all probabilities is certainty, by definition equal to unity.
- In an experiment, variable measurements can be treated statistically.
- The physics of the problem may well determine the statistical distribution to be used, for example the binomial or the Poisson distribution.
- Experimentally, the Poisson and normal distributions are of most use.
- Where experimental measurements are subject to an uncertain accuracy the result are usually assumed to be distributed normally.
- There is a 68 per cent confidence that a single measurement will lie within one standard deviation of a normal distribution.
- The standard error on the mean is taken to be the precision and is given by the ratio of the standard deviation divided by the square root of the total number of measurements.
- Each variable contributes to the total error on the final outcome of the experiment according to the sensitivity of the outcome to the errors on each variable.
- Where errors arise from resolution, the maximum error is taken as the error on the outcome. The maximum error is found by summing the individual contributions.
- A single counting experiment represents the best estimate of the mean in the Poisson distribution.
- The precision on a counting experiment is given by the square root of the total number of counts, so repetitive counts do not improve the precision.
- Where errors are determined from statistical considerations the standard deviation of the outcome is determined by the root-mean-square of the individual contributions.
- If co-variance is suspected, examination of the statistical distribution of the outcome is necessary.

- Minimising the squares is equivalent to finding the mean of a normal distribution.
- Least squares can be used to fit a function to experimental data.
- A scatter plot of the residuals will reveal systematic deviations of the data from the line of best fit.

▶ Notes

1 See under 'diffusion' in *McGraw-Hill Encyclopedia of Science and Technology*, 9th edn (New York: McGraw-Hill, 2002) vol. 5, p. 500.
2 See under 'Boltzmann statistics' in *McGraw-Hill Encyclopedia of Science and Technology*, 9th edn (New York: McGraw-Hill, 2002) vol. 3, pp. 198–200.
3 A. Yariv, *Quantum Electronics*, 3rd edn (New York: John Wiley & Sons, 1989) p. 576.

5　Theory in Physics

▶ A working view of physics

The role of theory in physics presents perhaps the toughest challenge to the majority of students. Theoretical physics involves not only learning and understanding difficult concepts, but also understanding the mathematical ideas in which these concepts are framed. Your learning in this area will depend very much on how you view physics, and as pointed out by Elby,[1] *'most students perceive learning physics deeply to be a significantly different activity from trying to do well in the course'*. The potential for conflict thus exists from the outset. Do you try to maximise your results by concentrating on those things most likely to bring success in exams at the expense of a deeper understanding, or do you concentrate on learning at the possible expense of your exam results?

Ideally one would imagine that if the theoretical aspects of a course have been understood and learnt, success in exams will follow, but not all students subscribe to this view. Research on physics education in the USA[2] has elucidated three typical views:

- learning physics is about retaining formulas and problem-solving algorithms;
- learning involves relating fundamental concepts to problem-solving techniques; and
- learning involves building one's own understanding.

It may be necessary at times to learn by rote and to cram, and to memorise formulae and problem-solving algorithms, but this has nothing to do with real learning or understanding. Kim and Pak[3] found that even after solving 1,000 traditional problems in physics, students do not overcome conceptual difficulties. Learning is a mysterious process, not always easily described or prescribed and learning theories and concepts is difficult. Your success will depend very much on what you do and how you approach your studies. Learning is an activity; you have to *do* something to learn. Learning represents a transformation in so far as your views should have changed to some extent after learning has taken place. You should then be better equipped to handle the challenges involved in your assessments because your answers will be based on understanding rather than repetition. Ideally there is no conflict between learning and doing well; by learning you might well improve your exam marks.

How you learn will also depend on how the physics is taught, and you have no control over this. Your understanding may be challenged or it may not. There may be an emphasis on traditional problem-solving, together with lectures in which you are simply given information to deal with as you can in your own time. Much of the information might contain detailed mathematical arguments in order to derive important results, and at other times the information may be very descriptive in order to emphasise the physics rather than the maths. If it is necessary for you to develop your own strategies for learning in order to advance your understanding of important concepts and ideas, it is important to appreciate how physics works so that you can understand what it is you are trying to achieve, which is to be able understand the theories you have encountered and also to be able to express ideas and concepts in the language of mathematics.

This chapter is therefore concerned to develop a working view of the process of physics itself rather than to provide a description of the areas of theoretical physics that currently lie at the forefront of our knowledge. This view is not intended to be philosophically rigorous, nor is it intended to put forward a definite view of what physics is, and by corollary what it is not. Rather, the intention is to raise the issues, point out some of the difficulties, and to explore the relationship between mathematics and physics. Ultimately, of course, the outcome of this is that you should be better placed to understand difficult concepts in physics, something that is not always achieved by the conventional approach of solving problems.

▶ Theoretical physics

Put simply, theoretical physics is the pursuit of knowledge of the physical world through the use of mathematics. The use of mathematics distinguishes physics from other sciences, such as chemistry or biology, but physics is *not* mathematics. Mathematics is concerned primarily with logical truth, irrespective of any physical reality that may be understood by that truth, but the object of physics is to match the physical world to the mathematical model. Physics and mathematics clearly enjoy a very close relationship, and some understanding of the nature of this relationship should help you to understand the role that mathematics plays in the formulation of theories.

Conventionally, undergraduate students learn how physics is done through practice; through the exposition of theories and the development of mathematical arguments in lectures and in textbooks. The practise of theorising or constructing mathematical arguments has been seen as something that is picked up as knowledge of physics is developed. However, this is a hit-and-miss affair. Some will pick it up before others, and some may never develop a clear understanding of the ideas. Hence there is a need to lay down some general principles. No doubt there will be disagreement among physicists as to the extent and scope, or even the content of these principles, but one thing that is abundantly clear is that even physicists cannot agree upon what it is they do. Some would argue, for example, that physics is about discerning the fundamental laws of the universe, which, taken at face value, seems hard to disagree with. As an operational statement, however, it contains a multitude of difficulties.

First among them is the notion of 'fundamental'. Take momentum, for example. In Newton's mechanics momentum is derived from the velocity, which is derived from the rate of change of position. In this one important respect, then, position is the most fundamental quantity. However, there are grounds for also thinking that momentum might be fundamental, even though it is derived from the position. After all, the law of conservation of momentum is one of the great conservation laws of physics. Not only this, but contrast Newton's formulation of mechanics with that of William Rowan Hamilton (1805–65). Hamilton was a genius in every sense of the word. Born in Dublin, he was fluent in 10 languages by the time he was 12 and taught himself mathematics at the age of 17. At the age of 21, while still an undergraduate, he was appointed Professor of Astronomy at Trinity College.

Hamilton reformulated Newton's laws of motion and in so doing developed an entirely different approach to mechanics in which the total energy of a particle is more important than knowledge of forces. The total energy is the sum of the potential and kinetic energy, commonly denoted by T, and related to the square of the momentum, p, by $T = p^2/2m$. This approach, which later formed the basis of Schrödinger's wave formulation of quantum mechanics, considers momentum to be fundamental. This, then, is the difficulty; to decide what constitutes the fundamental laws of the universe when what may appear fundamental varies from one perspective to another.

A second difficulty is that of the principle of reduction implicit in this idea. Reduction is the idea that a wide diversity of facts and phenomena can be reduced to a few principles, and is an important aspect of theoretical physics. Put the other way round, reduction to a few basic principles means that once these principals have been grasped the physicist can deal with a wide diversity of phenomena. Many physicists go further, though, and talk about the 'theory of everything', which is the reductionist ideal of a single grand principle. Whether such a theory can be developed or not is a matter of debate, but it is not the purpose of this chapter to engage in that debate. Much more important is what such a notion says about the way physics is done. One of the clearest findings to emerge from research in America into physics education is that students' views about the nature of physics influence the ways in which they learn, as demonstrated by the three different views on learning physics already described.

▶ Physics: separate subjects or interconnected ideas?

The question boils down to this; do students see physics as an interconnected web of ideas and concepts, or as separate facts to be learnt, memorised, and regurgitated? The notion that physics is a web of connected ideas is very common, but for the student this interconnectedness can be elusive. As taught at school, physics is presented as a series of discrete subjects; heat, light, sound, electricity, and so on. Even the development of the subject often appears this way. Many physicists, such as Newton, are noted for their work in different areas of physics; mechanics, discoveries in light, and even the invention of calculus, for which he shares the credit with the German, Gottfried Wilhelm Leibnitz (1646–1716). There is no suggestion that these were seen by Newton as different aspects of some underlying principle.

Within classical physics it is easy to see classical thermodynamics and mechanics as different, just as electricity and magnetism are different from fluid mechanics, for example. Occasionally, some developments in physics appear to unify seemingly separate ideas. Maxwell's theory on electromagnetism famously unifies both electrical and optical phenomena by showing that at root they are manifestations of the same basic phenomenon. At another level, however, they still appear separate. Geometrical optics does not make use of any overtly electromagnetic concepts, and to all intents and purposes it might as well be a different subject from the study of electrostatics. It is perfectly valid to ask, in what important respect are physics concepts related and interconnected?

Mathematics: the language of nature

The answer to the above is mathematics. Mathematics is the language physics uses to describe the natural world; it is the use of mathematics that by and large leads to common concepts. Take for example the mathematics of waves and vibrations. The same mathematical equations can be applied equally to electromagnetic waves, to acoustic waves, or quantum mechanical wave-functions. These are distinct and separate physical phenomena, but there is a commonality of ideas that unites them.

One of the inspiring features of mathematics is that physical phenomena have so often been predicted within the framework of a mathematical theory only to be subsequently verified experimentally. It would be very easy, therefore, to take mathematics as the reality, but the picture is not so straightforward. Paul Dirac, one of the pioneers of quantum mechanics, was a theoretician who trained as an engineer, during which time he learnt that mathematical approximations had a great deal of value. This experience informed his view of physics so that he sought to express physical quantities mathematically using simplifying approximations where possible. Dirac was quite unusual as a physicist in that he rarely tried to envisage the physical reality before expressing it mathematically, as most physicists would probably admit to doing, but instead concentrated solely on the mathematical description. The fact that he was content to use approximations does not suggest that Dirac thought that mathematics *is* reality, but that it *represents* reality.

Within the quantum world the correspondence between maths and physics is much looser. Just as there is more than one formulation of mechanics, there is more than one formulation of quantum mechanics. There are in fact nine formulations,[4] not all of which use the famous wave-function developed by Schrödinger. Even when Schrödinger developed his wave-mechanical formulation it was well-recognised that the wave-function had no physical meaning and was nothing more than a mathematical convenience. The existence of other formulations that make no reference to wave-functions simply confirms the fact. Mathematics, then, is a tool. Sometimes it is exact, and sometimes, as with Dirac, approximations are used.

▶ The development of mathematical physics

Mathematics and physics have developed alongside each other. Galileo was the first real scientist to use mathematics to describe natural phenomena.

Galileo was not a theoretician but an experimentalist, and although he is credited with the development of measurement as an essential aspect of experimental science, in fact he was not the first. Precise measurements were a characteristic feature of the astronomy of the time and pre-dated Galileo's developments in experimental physics.[5] Copernicus had published his highly mathematical account of the universe, and prior to this similarly detailed mathematical accounts of the Ptolemaic system had been published. However, astronomy was not regarded as a branch of physics. Astronomical science in this form was taught by professors of mathematics, and astronomers were regarded as mathematicians. On the other hand Natural scientists, what we would now call chemists and physicists, were regarded as philosophers, and an entirely different discipline based on Aristotle's book *On The Heavens* was taught by professors of philosophy.

The predominance of Aristotelian philosophy can be traced to the humanist movement that arose in the two centuries preceding Galileo. The humanists believed that all knowledge lay with the ancients, and therefore concentrated on obtaining Greek manuscripts and translating them as accurately as possible. Translations of many works already existed at that time, but these had come via intermediate languages such as Arabic, and were considered to be distortions of the truth. Hence, at this time the world view was dominated by the qualitative science of Aristotle, which was not based on any quantitative notions; gravitational attraction, for example, occurred because the earth was the heaviest element and naturally took its place at the centre of the universe, where other matter was quite naturally attracted to it. Galileo differed from those who came before, including the astronomer-mathematicians, because he investigated the world around him. His innovation was to replace the qualitative approach with a quantitative approach, and in the process he began to discover new phenomena about the world.

Experimentation and quantification
As so often with historical figures, this simple account overlooks the context in which Galileo was working. At the time of Galileo's birth there had already arisen the strong view in some circles that Aristotle was inadequate, so the time was right for such a figure to emerge with a different approach. This said, Galileo's ideas about experimentation were not accepted easily by the philosophers of the time. Aristotelian science was based on pure reason; it stood to reason, for example, that heavy bodies fell faster than light bodies. Therefore it must be so, and accepting this there is no need to perform an experiment. Galileo, on the other hand, used mathematics

to quantify how far an object would fall in a given time, and then sought to verify this prediction using experimentation.

It follows that the experimenter has to take care to devise an experiment that will yield a 'true' measurement of the phenomenon under investigation. For example, it was not until Galileo made a series of precise measurements of motion down inclined planes that he was able to deduce that the distance travelled depended on the square of the time over which the motion occurred. If these measurements were to form the basis of a theory of motion under a gravitational force, the idea being that the theory had to reproduce this square-law behaviour, of course it was necessary to have confidence in the measurements. The very fact of making such precise measurements raised several new difficulties, first among which was the notion of units. Galileo's measurement unit was about a millimetre in length, and the accuracy was limited to half of this unit, which is the limitation of the resolution of the eye. Galileo recognised the difficulties involved in the measurement and was therefore not concerned to achieve an exact correspondence between mathematics and experiment, though of course he took as much care as he could.

The truth remains, though, that without experimentation there could be no quantification, and without quantification there could be no mathematics. For example, experiments on pendula indicated that the period was constant irrespective of the angle through which a pendulum swung, so Galileo had to confront the idea that the motion changed in such a way to compensate for changes in distance. This led to the notion of continuous changes in motion, which was not something appreciated by philosophers of the time. According to Drake, Aristotle had defined 'equal velocity' and 'greater velocity' and medieval philosophers were therefore content with these notions. In consequence the philosophers were unable to resolve problems such as that presented by Zeno's paradox (Zeno of Elea lived around 450 BC) in which a fast runner (Achilles) setting off after a slow runner (a tortoise) would never be able to catch up with the slow runner. By the time the faster reaches the position occupied by the slow runner at the time the faster started, the slow runner will have moved on. By this reasoning the faster runner will never catch the slower because the slower runner will always have moved on just that little bit further.

▶ The problem with 'pure' reason

It is not always so easy today to appreciate how seductive, and false, the notion of 'pure' – by which is meant Aristotelian – reason can be. From an early age we are introduced to the idea of experimentation, of mathematical analysis and prediction, and the idea that theories are testable in a real and meaningful way, so it is hard to imagine a time when the world was not seen in this way. Consider then, the following question:[6] suppose two books of differing weight, but with identical covers are 'kicked' such that they have the same starting velocity; which travels furthest? Ask this of somebody with little or no knowledge of mechanics and nothing but 'pure reason' to rely upon, and the answer given will probably be wrong; namely, that the heaviest travels furthest.

This is very similar to the question answered by Galileo in his famous experiment conducted from the leaning tower at Pisa: two bodies of different weight are dropped at the same time; which hits the ground first? The answer, as we know, is neither; they both hit the ground at the same time, but to the natural philosophers of the time it was obvious that the heaviest should travel fastest. Obvious perhaps, but no less incorrect for it; so how is it, then, that reason failed the philosophers of the time? Just as reason led to the wrong answers in Renaissance times, it would be wrong to conclude from it that reason is not important. Of course it is; a logical argument, whether conducted in philosophy or physics, must be based on reason. Where the medieval philosophers failed was in not recognising the assumptions upon which the arguments were based. All arguments have a starting point. These may be facts, or they may be assumptions, and Aristotelian reasoning was based on the assumption that the Aristotelian view of the world was correct. It is understandable that such a mistake might have been made; so ingrained was the view that Aristotle's philosophy expressed the truth about the world that it was not questioned. Galileo made no such assumption and therefore experimented. Thus he appealed to nature to provide the facts, and having determined the facts was able to construct sensible arguments to explain and predict natural phenomena.

▶ Mathematics as symbolic logic

We know from experience that the argument that Achilles will never catch the tortoise is false, but it is not so easy to see where the flaw in the argument lies. It is this conflict between the seemingly correct argument derived from reason and our experience that leads to the paradox. From the foregoing you might suppose that the answer lies in examining the assumptions of the argument. Indeed, the flaw lies here, in the notion of infinity. The argument to resolve the paradox must not only be logical, therefore, but it must also be mathematical in nature. The paradox arises because the motion is assumed to occur in little jumps and the slow runner always remains one jump ahead. As there must be an infinite number of such jumps the slower can never be caught. However, if the interval of space over which the race is run is divided infinitely, the sum of all the divisions must equal the total distance of the race. Mathematically this is known as a convergent series. Some series diverge; that is, the sum of an infinite number of terms tends to infinity, but others such as this tend to a finite number. Therefore, even if the motion occurs in jumps, ascribing a definite velocity to the runner means that the sum of all the times taken to cover the infinity of sub-divisions is just the time taken to cover the total distance. Simple; the fastest runner wins.

It is possible to approach the problem differently, using the notion of continuous velocity. The concept of an instantaneous velocity that could be measured and that could be changed during motion was invented by Galileo, and it took him many years to develop the mathematical tools necessary to develop the notion fully. Using such notions as velocity and acceleration it is possible to calculate how long it should take a runner to cover a certain distance and therefore whether one runner will overtake another if they start at different times and travel at different speeds. It is even possible to calculate the point at which one passes the other. Viewed in this way, the paradox does not arise at all. The reasoning is that the faster runner will take a shorter time to cover the distance and therefore must win. The argument is entirely logical, but it is based on the symbolic language of mathematics, which goes beyond mere arithmetic. Calculating whether one runner overtakes another is simply a matter of arithmetic and if that were all that could be achieved the mathematical argument, while useful, would be trivial. The power of mathematics is that it allows general arguments to be developed from which the specific can be derived. Mathematics is therefore *symbolic* logic.

The question posed about which book travels furthest can be answered by a similar appeal to the symbolic logic of maths. The resistance to motion is proportional to the normal reaction, which is proportional to the weight of the book (Figure 5.1). The resistive force is also related to the properties of the surface, but since these are explicitly made equal they have no influence in this argument. The resistance to the motion is thus constant and independent of the velocity.

The work done by the resistive force is equal to the product of the force times the distance over which the force acts, and must be equal to the initial kinetic energy of the body, which depends on the mass and the square of the velocity. As the starting velocities are equal, only the mass

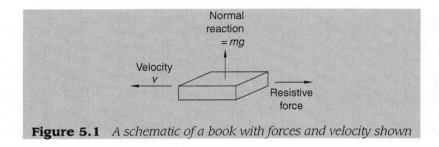

Figure 5.1 *A schematic of a book with forces and velocity shown*

need be considered. Thus the mass appears in the work done and the kinetic energy, and therefore cancels out, leading to the conclusion that the distance travelled is independent of the mass *in the circumstances as described*. Change the circumstances, of course, and the answer changes. If the bodies are given equal kinetic energy rather than equal initial velocity, the lighter book will travel furthest. Even this might contradict our first thoughts.

▶ Dialectical physics

The question posed above on the two books is a form of dialectical argument. Dialectics is the name given to the pursuit of truth through discussion, and was a favoured technique of Greek philosophers. In particular, discussion allows the conflicts resulting from opposing views to be brought to the fore, and from this confusion an understanding of the truth can emerge. In the problem above, our initial supposition about the answer is in conflict with the facts, but through discussion and the formulation of a logical argument based on the symbols of mathematics, we can resolve the conflict. Of course, our initial supposition about the answer may not be wrong, but it doesn't follow that the thinking behind the supposition is correct. Even with a correct first answer there can be value in discussion.

Physics is not always taught in this dialectical manner. If this problem of which book travels furthest were to be tackled within a lecture course, then more than likely the issue would have been approached by setting up the equations for the work done in stopping the body and showing that under certain circumstances the distance travelled is independent of the mass. Since the mathematics expresses known physical laws about the mechanics of motion the arguments are entirely logical and easy to follow, but there is no guarantee that these concepts will have been learned. Learning is an activity; you have to *do* something to learn and simply listening to an argument is not enough. The element of surprise that comes from discovering that initial thoughts are wrong will be missing. Being confronted by our confusion and working through to the conclusions is an activity that results in real learning such that the knowledge is internalised and we understand. If confronted with a different problem we would have little difficulty in using the principles thus understood to arrive at the correct answer.

The dialectical approach is one way of taking control of your own learning. You have no control over the manner in which information is presented to you, but by recognising that confusion indicates not only a lack of understanding but also that an opportunity exists to learn, you can ensure that such confusion, when it arises, can be used to your advantage. Ultimately the argument has to be expressed in mathematical form. Intuition is sometimes a useful guide, but at other times it can be misleading, and as with the case of Zeno's paradox, a mathematical argument settles what might otherwise be no more than a clever dispute over words.

▶ Mathematics in physics

The relationship between mathematics and physics can now be described. The physical world can be described by mathematical relationships. Sometimes, as in the case of Galileo, these relationships are derived from experiment. The purpose of theoretical physics is to develop these mathematical relationships and use them to determine the nature of the physical world.

Naturally, this requires a detailed knowledge of mathematical techniques. Sometimes the detailed arguments are so mathematical that theoretical physics is indistinguishable from applied mathematics. However, the mathematics is not the most important part of this process. Mathematical arguments, being based on logic, are true whether a physical reality is represented in the argument or not, and can be developed by a mathematician with no interest in physics if necessary. Where the physics is important is primarily at the beginning and the end of this process. First, the physical reality has to be expressed mathematically, and after the detailed mathematics has been developed the resulting mathematical expressions have to be interpreted in the physical world (Figure 5.2). The real ability of the theoretician, therefore, lies in the ability to relate the mathematics to the physics.

An example will help at this point. Consider the vibrations of a mass on a spring, something very familiar. We know that the acceleration a depends

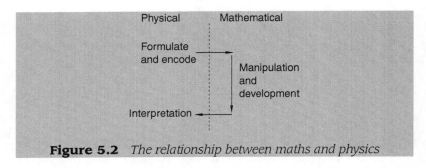

Figure 5.2 *The relationship between maths and physics*

directly on the force applied and inversely on the mass, according to Newton's second law. Thus:

$$F = ma \tag{5.1}$$

We can write the acceleration as the rate of change of velocity, dv/dx, which in turn can be written as the rate of change of position, dx/dt, in one dimension x. Hence:

$$a = \frac{d^2x}{dt^2} \tag{5.2}$$

So far this is general and not specific to a mass on a spring. In order to describe this particular problem we need to describe the force, which we know from Hooke's law to be proportional to the extension of the spring and directed toward the centre. Putting the equilibrium position of the spring as $x = 0$, the force can be written as:

$$F = -k \cdot x \tag{5.3}$$

where k is a constant of proportionality, known as the spring constant or sometimes the stiffness. It can be found experimentally by the simple expedient of hanging different masses from the spring and recording the final position. At this point there is no net motion so the downward force exerted by the mass, that is the weight, must equal the upward force exerted by the spring. A plot of spring extension against mass therefore yields k.

The force is negative because it is directed against the motion. As the spring extends the force increases but acts to direct the mass towards the centre of the motion; that is the rest position of the spring. Inserting this force into Newton's second law yields the equation of motion for the mass–spring system, that is:

$$m\frac{d^2x}{dt^2} = -k \cdot x \tag{5.4}$$

How this equation is solved is not of interest here, though the maths of simple harmonic motion, as the solution of this equation is known, is described in greater detail in the next chapter. The important point here is that this equation now expresses the essential physics of the problem in mathematical form. The forces acting have been described, and the resulting acceleration has been incorporated into the equation. This constitutes the first part of the process illustrated in Figure 5.2. It remains to solve the equation to find the particular form that the motion takes, and though it is relatively easy it won't be done here. The solution is an exercise in mathematics, so we'll simply accept the outcome of the mathematics and move swiftly on to look at the physics of the solution. This is the last part of the process, that of interpreting the result.

The solution to this equation is:

$$x = A \sin(\omega t) \hspace{3cm} (5.5)$$

Here A is a constant called the amplitude. The interpretation of this equation is based on the following:

- ωt must have the units of an angle;
- the angle increases with time;
- the sine function is cyclic, varying between $+1$ and -1;
- the extent of the motion therefore varies between $+A$ and $-A$;
- the period of the oscillation, that is the time for one complete cycle from say $+A$ through 0 to $-A$ and back again to $+A$ is $2\pi/\omega$;
- this motion will continue indefinitely.

▶ Hidden assumptions

The last point is very important in so far as it represents a situation that is unphysical. The description of an oscillation in terms of a sine function is exact in many cases and a very good approximation in others, but experience tells us that if we set something oscillating then eventually it will stop. This solution contains no such mechanism, but the fault does not lie with the solution. This is the important point about the mathematics. The solution of this equation is exact, so if the solution does not describe the physical reality it is the original equation that is in error.

The error lies in the *assumption* that the force exerted by the spring is the only significant force acting. You may question this; we haven't explicitly assumed anything, so why now does it appear to be an assumption of the model? It is not an explicit assumption in so far as we didn't state that we do not assume any other forces to be acting, but the very fact of excluding them is equivalent to making such an assumption. This can make theoretical physics very difficult, because in setting up the equations for a system in order to frame it in mathematical terms all sorts of hidden and unrecognised assumptions are made. It is sometimes necessary to go back and re-examine them, but of course the fact that they are hidden makes this very difficult. Nonetheless, in the present case we recognise the assumption and recognise that it is deficient. Let us therefore put a resistive force into the equation.

The form of the resistive force involves making another assumption. Many textbooks simply state that the resistive force is proportional to the velocity, which is true in many cases but not all. The problem of which book travels furthest presented earlier contained a resistive force that did not behave in this way, so there is no guarantee that all resistive forces are characterised by a dependence on the velocity. Nonetheless, we can make the assumption that this is the nature of the force, look at the solution and see if it bears any relation to reality. If not, it is necessary to adopt a different form of force. Therefore we write:

$$m\frac{d^2x}{dt^2} = -k \cdot x - \gamma \frac{dx}{dt} \tag{5.6}$$

Again the negative sign in front of the resistive force indicates a force acting against the motion. This has the solution,

$$x = Ae^{-\gamma t/2} \sin(\omega t) \tag{5.7}$$

which is not always written in this form. Very often oscillatory motion is expressed in terms of a complex exponential, but as this has not yet been described, and you may not be familiar with it, let's stick with the sine function. In fact the sine function can be derived from the complex exponential so is not actually very different mathematically. The difference here is that the amplitude now contains a term that decreases as time progresses, and the rate of decrease depends on the strength of the resistive force via the constant γ. Many physical systems are found to be described by an equation of this sort, but again it must be emphasised, if it is found that the decay is not exponential as shown here it is the original expression of the nature of the resistive force that is wrong, not the solution.

It is possible to go further with this particular example. You may well be aware of systems that do in fact appear to oscillate indefinitely, and so you may think that the original solution is correct. However, it is clear that the physical causes may well be different. It is one thing to set an oscillator in motion and have it continue indefinitely, as in the first case, but it is quite another to have an oscillator driven in some way. Such an oscillation will last as long as the driving force lasts, and is called a 'forced' oscillation. The most common driver is a force that is sinusoidal itself, such that:

$$m\frac{d^2x}{dt^2} = F_0\cos(\omega t) - k \cdot x - \gamma\frac{dx}{dt} \tag{5.8}$$

where the term in F_0 is the driving force. Physically this may correspond to a variety of situations, for example an ion in a crystal subjected to a steady electromagnetic wave. The electric field of the wave oscillates at the frequency of the wave, and a force is exerted by the field on the ion simply because the ion is charged. The forces on an ion are actually quite complicated to describe, but it is clear that the ions are held fairly strongly in certain positions. Otherwise it would be easy to either stretch or compress a crystalline solid, and experience tells us this is not the case. Thus an ion is held in a particular position, surrounded by its neighbours, and if something happens to move the ion from this position forces act to restore it to the centre. The situation is more complicated than a mass on a spring for the simple reason that all the ions can oscillate whereas with a spring there is only one oscillation to consider. If we were to ignore the oscillations of the other ions and concentrate on only one, we would find that it can be adequately described by an equation such as that given above. The solution is again given by equation (5.5):

$$x = A\sin(\omega t)$$

but with the amplitude being given by,

$$A = \frac{F_0\sqrt{\left(\omega_0^2 - \omega^2\right)^2 + (\gamma\omega)^2}}{m\left(\omega_0^2 - \omega^2\right)^2 + (\gamma\omega)^2} \tag{5.9}$$

Again this is expressed slightly differently from the form commonly found in most texts. The amplitude is usually expressed as a complex number, but I have taken it a stage further to eliminate this aspect. Before this can be interpreted, it is necessary to define ω_0, as it appears here. Had we been developing the solution to the equation from the beginning it would have been clear what this quantity represents, but with the solution being presented as a fact it is necessary to explain its origin. Obviously it is a frequency of some sort, but it is not the frequency of the oscillation. Instead it is a fixed frequency, called the natural frequency of vibration of the system. This is the frequency that would in fact be observed if an undamped, unforced oscillator were to be set in motion and left to oscillate.

The essential physics is contained within the expression for the amplitude and the following can be deduced:

- For a low frequency $\omega \ll \omega_0$ the term $(\omega_0 - \omega)$ approximates to ω_0. As $\gamma\omega$ is likely to be much smaller than ω_0 – a large value of γ implies a large resistive force, but there must be a physical limit otherwise the mass will not move in the first place – this term can be ignored. The amplitude therefore approximates to:

$$A = \frac{F_0}{m\omega_0^2} \tag{5.10}$$

Under these circumstances the amplitude is proportional to the amplitude of the driving force, which is only to be expected.

- As ω approaches ω_0, the amplitude increases because $(\omega_0 - \omega)$ approaches zero. At the point $\omega_0 = \omega$ the amplitude becomes:

$$A = \frac{F_0}{m\gamma\omega} \tag{5.11}$$

- This is the maximum value of the amplitude, as will become clear below. The important point is that the presence of the resistive force limits the magnitude of the amplitude, as can be seen by setting $\gamma = 0$. It doesn't matter whether this substitution is made before ω is set equal to ω_0 or after; the amplitude becomes infinite. This is known as resonance, hence the designation of ω_0 as the resonant frequency. Therefore, the greater the resistance the smaller the amplitude at resonance.
- For high frequencies such that $\omega_0 \ll \omega$ the amplitude approximates to:

$$A = \frac{F_0}{m\omega^2} \tag{5.12}$$

The amplitude therefore decays to zero as the frequency increases.

▶ Qualitative and quantitative predictions

These deductions can be combined to produce the schematic variation of the amplitude as a function of frequency shown in Figure 5.3. At low frequencies the amplitude is independent of frequency, rises to a maximum at the resonant frequency, and decreases to zero at high frequency. This is a qualitative description of the behaviour. Many examples can be found in the physical world that will behave in like manner, but the question then is, does this theory apply to these systems?

The answer lies not in the generality but in the detail. The solution allows far more than a general description of the behaviour of the model as a key

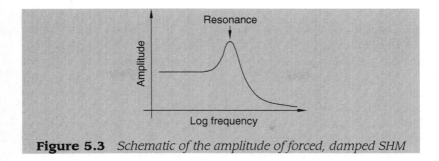

Figure 5.3 *Schematic of the amplitude of forced, damped SHM*

parameter is varied, it allows for an exact calculation of the effect, in this case the amplitude. In any well controlled experiment the size of the driving force F_0 will be known, as will the mass, the resistive force, and the driving frequency. It is therefore possible to calculate *exactly* the amplitude at any given frequency, and hence compare this value with experimentally measured values. This implies, of course, that the experiments are conducted with sufficient accuracy for the comparison to be meaningful (see Chapter 3).

▶ Hidden implications

Just as there are hidden assumptions, there are also hidden implications. Both have to be identified and neither is easy. Open-mindedness is required, as well as logic and reason. It is particularly important to be able to identify an effect with a cause; for example the restriction on the amplitude at resonance due to the presence of the resistance. Resistance is the cause, a finite amplitude is the effect.

A hidden implication of the foregoing example is that energy is absorbed from the driving source at resonance. It does not strictly emerge from the solution, but can be argued as a logical consequence, because the energy of the oscillator is maximised at this point. This energy has to be supplied from somewhere, and it can only be from the driving force. Using this understanding, it is possible now to appreciate why crystalline materials absorb infra-red radiation at certain frequencies by exciting vibrations of the lattice; when the frequency of the incident radiation is equal to the natural frequency of vibration of the lattice a resonance occurs. It is also possible to understand why mechanical systems vibrate and exhibit resonance, and why supplying some sort of resistance can reduce the magnitude of the vibration at resonance because energy is dissipated in overcoming the resistance.

There are occasionally historical examples of the inability to appreciate the implications of a theory. Felix Bloch, while a research student under Heisenberg, solved the tricky quantum mechanical problem of a particle travelling within a periodic potential. This was an important development at the time, because the existing theory of conduction in metals ignored the fact that a solid consisted of atoms. It would be reasonable to suppose that an electron should be affected by atoms, or rather by the periodic fluctuations in the potential energy arising from the regular spacing of the atoms, but by ignoring them the mechanism for electrical resistance had effectively been assumed out of previous models. There was nothing off which the electrons could be scattered. The result was the so-called 'free-electron' model of the atom, which despite its failure to account for electrical resistance, was in fact a very successful model in many respects. Bloch's solution showed that if a crystal is perfect there should be no resistance. The atoms are effectively 'invisible' in a perfect crystal but any deviations from perfection cause the electrons to scatter. Story has it that when Bloch solved the problem he did not recognise the importance of his finding and it was left to his supervisor, Heisenberg, to recognise the result.

Whether true or not – I read the story many years ago and cannot recall the source – the story demonstrates the importance of critical thinking. Within the context of theoretical physics it is a vital skill, especially if the outcome of a theory is not what is expected. It is necessary then to assess the worth of the theory by looking at the assumptions and also whether there are testable predictions. An example of the latter can be seen in the examples of oscillatory motion; the amplitude of a damped oscillator decays exponentially with time and the amplitude of a forced damped oscillator has a characteristic behaviour as a function of frequency that can be calculated exactly. These are *predictions*. If the theory is right, experiments performed under the conditions corresponding to the model should yield results in accordance with the mathematics. If they do not, something is amiss. Alternatively, if a system behaves as described it means that the forces behave in the manner formulated in the model. The predictive power of mathematical models is one of the marvels of physics.

Skill in mathematics

The foregoing discussion has demonstrated very clearly how it is possible to express a physical situation in mathematical terms, and then, once a solution to the equations has been obtained, understand the physical implications of the model. The concentration has been solely on these two aspects and the mathematical techniques for developing the solution have been ignored. You should not conclude from this that maths is not important. It is. Skill in mathematics is not only needed to develop the solutions, it is also a vital aspect of the formulation of the problem.

For a significant number of physics students, skill in mathematics is hard to acquire. The logical arguments of mathematics are often easy to follow in a lecture theatre, but when it comes to applying the techniques to solve mathematical problems difficulties are encountered. If this applies to you, the only way to acquire the necessary techniques is to practise, practise, and practise some more. However, skill in mathematics is only one aspect of success in theoretical physics.

▶ The scientific mind

The critical thinking skills described above are part of the scientific approach, and it is worth examining the historical development of the scientific mind in order to see what it can tell us about studying physics. By the term 'scientific mind' is meant principally the processes of deductive and inductive reasoning that characterise the approach of many scientists even today. The origins of this scientific philosophy can be seen in the Renaissance but it is possible to discern the development of distinctive views on physics during subsequent decades that help us to understand and appreciate the nature of physics.

We start with Galileo, who is commonly credited with the beginnings of scientific investigation as we know it today, having pioneered developments in mechanics, optics, and astronomy. Galileo's telescope was the first used for astronomical observation, allowing him to see that Jupiter had moons and that Venus showed phases similar to those of the moon. However, it was in fact Francis Bacon (1561–1626), born just three years before Galileo, who can take credit for providing much of the philosophical basis for our modern scientific method. Bacon believed that science could greatly improve the human condition and further believed in the right of man to dominate nature. It seems extraordinarily egotistical in these days of environmental awareness to think that the domination of nature is a legitimate aim of science, and indeed it seems to have no place in modern notions of physics, which is about *understanding* the world rather than dominating it. However, think of some of the great engineering achievements of the last two centuries and the idea of dominion over nature can be seen clearly.

Induction, deduction and reduction

Bacon was an inductivist. He believed that enough observations of a particular phenomenon would allow an observer to induce the fundamental principles involved. Bacon was essentially an experimentalist, but not the quantitative experimentalist that was Galileo. Rather his work was qualitative. Contemporaneously, René Descartes (1596–1650) proposed a different approach to the development of science, believing instead that the basic ruling principles of nature were accessible through reason and mathematical logic. Descartes invented cartesian geometry, which allows geometrical figures to be expressed as algebraic equations, and the use of cartesian coordinates is one of his lasting legacies. Descartes was a *reductionist* who believed mathematical analysis allowed physical laws to be deduced, rather than induced, and in consequence a wide-ranging variety of physical phenomena can be reduced to a few basic principles.

It is easy to regard these two approaches, Bacon's qualitative inductive method and Descartes quantitative deductive/reductive method, as opposing. However, replace Bacon's qualitative empiricism with Galileo's quantitative methods while retaining his inductivist views, and you have within both methods the core of modern scientific method. The history of physics is thus the discovery of phenomena and laws through both experimentation and mathematical analysis; theory and experiment not in opposition but complementing each other.

Practical deduction and induction

All three processes of induction, deduction and reduction are related in the modern approach to physics, but deduction and reduction are by far the most important. Induction as a philosophy of science has severe limitations. How many observations should be made before a law can be induced? Just because something has happened a thousand times does not mean that it will always happen. Induction is of limited applicability in physics, but that is not to say that it has always been the case. At the time of Bacon and Galileo (Newton came a short time later) very little was known about the physical world. In fact, knowledge of the natural world was regarded as a branch of magic; not the spells and sorcery that everyone associates with the word today, but an expression of the wonder and the hidden secrets of the natural world that were seemingly beyond explanation. It was reckoned to be part of education to be familiar with magic.[7] Knowledge of the physical world was so limited that it may have seemed perfectly reasonable to induce fundamental laws from a few observations, but today we would regard such an approach with suspicion. We would recognise an observation as one fact to be considered among many.

We owe this situation to the development of deduction; the formulation of mathematical arguments from which physical laws can be deduced. The essential problem with deduction, though, is that if the initial assumptions and the consequent formulation of the problem are not correct, then no matter how correct the mathematics the conclusion is bound to be in error. Deduction does not work unless it is backed up by experimental observation which provides the initial facts and the test for the conclusion. Occasionally, where theoretical developments are lacking, experimental observations take on a greater importance as they constitute our primary knowledge of the field, but the fact that a theory had not been developed would more than likely lead us to conclude that our observations are only part of an incomplete picture. Experimental observation is thus very important but not in the inductivist sense.

Reduction and the theory of everything

The reductionism of Descartes is limited. It occurs naturally as a consequence of deduction, but some physicists would go further. In the words of the mathematical physicist Laplace (Pierre-Simon Laplace, 1749–1827):

> I have sought to establish that the phenomena of nature can be reduced in the last analysis to actions at a distance between molecule and molecule, and that the consideration of these actions must serve as the basis of the mathematical theory of these phenomena.

In 1819, towards the end of his life, he wrote:

> If an intelligence, for a given instant, recognises all the forces which animate Nature, and the respective positions of all things which compose it, and if that intelligence is sufficiently vast to subject these data to analysis, it will comprehend in one formula the movements of the largest bodies of the universe as well as those of the minutest atom; nothing will be uncertain to it, and the future as well as the past will be present to its vision.

This is the ultimate in reductionism; everything can be reduced to a single principle, the theory of everything. This view is typical of many great physicists, including people such as Einstein, but it is questionable not only whether this type of reductionism can tell us anything useful about the process of physics, but also whether it is a tenable view. Are we to regard current theories as no more than approximations until a better theory comes along? If we are honest and open-minded about it this might not be such a problem, but if it is used as the basis for rejecting a workable theory then we are ill-served by the philosophy. How, then, did such a view arise and what might constitute a better approach to physics?

Determinism and reduction

Laplace's views were expressed long before the advent of quantum mechanics and before Poincaré (Henri Poincaré, 1854–1912) predicted the chaotic behaviour that sprang to prominence at the end of the twentieth century. Both of these developments have changed the way we view the world. In saying that, *'future as well as the past will be present to its vision'*, Laplace was actually referring to the developments in astronomy, which, as we know, had been studied in great detail since before Galileo and for which detailed mathematical models had been developed. Judging by the development of astronomy the world seemed solidly deterministic. It is ironic, then, that it is in relation to an astronomical problem that Poincaré first proposed the turbulent and non-deterministic behaviour that we call 'chaotic' today.

The reductionist view owes much to determinism. It seems perfectly sensible to suppose that if the world operates along well-defined lines such that we can predict any outcome given sufficient knowledge, then clearly, if we are given this knowledge we will know everything and see how everything works together. We will have the theory of everything. The fact that the world is not deterministic – and not just the quantum but also the macroscopic world – implies very strongly that such a theory is probably not attainable because there are realms not susceptible to prediction. Ultimate reduction appears to be a wish rather than a realistic aim. It certainly does not appear to be a practical view.

▶ Hierarchies in physics

Even today, however, there are physicists who subscribe to this form of reductionism. Quite possibly it is connected with the way that physics has developed, aided by technology. We have gone from the large to the small, down to the atomic in the 1930s to the sub-atomic, the sub-nuclear, and beyond. Perhaps it is perfectly natural to imagine that there must be a limit, rather like hitting rock when digging a hole, and that once we have reached the 'bottom' we'll understand ultimately how matter is put together. Along the way to these discoveries various forces have been unified and are seen to be different aspects of one phenomenon, which is clearly an example of reduction. It is reasonable to suppose that the ultimate result of these endeavours will be the reductionist's dream of an ultimate theory. However, turn the argument around and ask, suppose we have such a theory how much of the physical world can it really explain? Ask such a physicist if the ultimate theory of matter could be used to explain how a transistor worked, for example, and the answer would more than likely be 'no'.

This leads directly to the concept of hierarchies in physics. Even if we understand matter and the stuff of the universe in its most basic and fundamental form, we can't really call it reduction in the manner proposed by Laplace if, as we move up the scale to the sub-nuclear, the nuclear, the atomic, the solid state, the gaseous and so on we have to develop new concepts and new approximations in order to deal with these phenomena. For that is the situation that faces us today. In atomic physics we are not concerned with what happens at the nuclear level but at the electronic and there is a range of theories and concepts that describe these phenomena. When these atoms are combined to form a solid we are faced with the fact of interacting electrons that behave differently from electrons in an atom, and new concepts have to be developed to explain phenomena such as electrical conduction.

These hierarchies are not entirely separate. There are many common concepts that link them, but there are also clear differences that separate them. In answer to an earlier question, is physics made up of separate ideas or a web of interconnected concepts, the answer is, 'both'. It is perfectly legitimate to see in physics a separateness between electromagnetic theory and geometric optics, for example; they form different levels within the hierarchy. Electromagnetic theory is at a lower, more fundamental, level. Quite possibly it could be used to explain most of the phenomena for which geometric optics is used. I don't know that this is the case, I haven't tried it, but I have no doubt that if it were possible the formulations and calculations would be very complicated and certainly less convenient. Geometric optics therefore performs a particular function that electromagnetic theory does not, though there are clearly many concepts, such as light-waves, refractive index and so on, that link the two.

A working view of physics

It is now possible to put together a working view of physics as follows:

- Nature is described by mathematics.
- Representation of physical phenomena by mathematical equations allows new laws to be deduced.
- The process of deduction is informed and tested against experimental observation.
- Deduction can unify diverse phenomena and is a form of reduction.

- Concepts in physics can be organised into hierarchies.
- Hierarchies can appear distinct but are linked by common ideas.
- Reduction is local and takes place certainly within hierarchies and possibly between them, but a theory of everything is probably not possible.

This is a working idea of the process of physics. We can see in here the relationship between maths and physics, the principles of deducing laws from diverse facts, and how to use experimentation in combination with theoretical developments. The notion of hierarchies is developed but the hierarchies themselves are deliberately not defined. It is up to you to see the common threads and connections, but where concepts appear distinct and unconnected from others it may well be that there is a good reason. Much depends, of course, on your level of understanding. At a simple level different areas will appear very definitely separated, but when you begin to appreciate the linking ideas they may be but different manifestations of the same thing and not separate at all. However, develop your knowledge further and begin to specialise and the differences between subjects will become apparent again.

▶ Physics and the philosophy of science

The philosophy of science has been discussed in Chapter 1, and effectively dismissed as not saying much of use to the working physicist. This is largely true but not completely. Much of the previous discussion has been concerned with the philosophy of how we understand and do physics, and processes such as induction and deduction will be familiar to philosophers of science. However, our philosophy is specific to physics. One of the problems, it seems to me, with the philosophy of science is that science is treated as a homogeneous activity when in fact different sciences use different methods. One of the distinguishing features of physics is the close relationship with mathematics that has rendered induction virtually meaningless and elevated deduction to a position of prominence. This might not be true across the sciences in general, where a closer reliance on observation may require different methods. However, I am not concerned with other sciences.

Where I personally take issue with the philosophy of science is in the inevitable notion that if we can identify the way science works and develops then as scientists we should organise our activities around these aims. This ignores the fact that physics is studied and researched out of a love for the subject and a fascination with the mysteries and, lets be honest, beauty of nature. Take Popper's view on falsification, for instance. Karl Popper is one of two philosophers of science who, if he is not widely read by physicists, is at least known to them. Thomas Kuhn is the other. Falsification has it that it is not possible to prove a theory right but it is possible to find circumstances in which it is wrong. Although the idea of falsification is basically correct, I know of no physicists who actively spend their time trying to falsify theories. On the contrary, we spend our time trying to do just the opposite; to prove our theories.

Falsification and paradigms

Physics is constructive whereas falsification is destructive. As physicists we want to use experimental observations as the raw material upon which mathematical theories can be based, from which conclusions – sometimes laws – can be deduced. We then use the predictive power of the theory to design further experiments to test the predictions. If the predictions don't match the reality we re-examine the assumptions and start again. This is actually a different process from falsification. Where falsification is important is in the overthrow of existing theories. Sooner or later we arrive at a point where the weight of evidence, experimental or theoretical, suggests that a particular view is false and we then derive another. Falsification serves a very important purpose; it reminds us that theories are only theories, and that no matter how correct they appear to be there may be circumstances just around the corner that will reveal to us their limitations.

There are occasions when the development of a new theory significantly changes our view of the world. Thomas Kuhn therefore developed the idea of a paradigm. This is not an easy concept to grasp. According to the Oxford English Dictionary a paradigm is a pattern, and in the context of a theory it is taken to be the prevailing idea of the time. This is actually a very useful idea that fits very well with the notion of hierarchies of distinct fields. As described, there are connecting ideas but some ideas don't merely link the hierarchies, they seem to underlie the structure. Once upon a time it was classical mechanics. In the pre-quantum era, for example, classical mechanics described the motions of atom in a gas from which kinetic theory and later statistical mechanics were developed, as well as describing the motion of the planets and the stars, and led Laplace to his reduction view. Today quantum mechanics has taken its place and can be found in the physics of atomic systems, solid-state systems, nuclear systems, optical systems, and particle physics. It is even relevant to the theory of black holes. Quantum mechanics isn't just a linking idea, therefore, it is the sub-structure on which the edifice of modern physics is built. In Kuhn's language, quantum mechanics is the paradigm.

▶ A new theory or a different approach?

The preceding discussion on the relationship between mathematics and physics has concentrated naturally on the development of theories. Where a theory is seen to be deficient, the natural assumption is that a new theory needs to be developed. This is not always the case, as the following example will show.

Consider again oscillatory motion, but now under the action of a non-linear force, for example a pendulum. A pendulum is constrained to travel in a circular trajectory by the fixed length of the pendulum. We know that the velocity of interest is directed at a tangent to the radius and therefore so is the acceleration. The force of interest must also act along the tangent. As the only force acting is gravity, we can write:

$$F = -mg \cdot \sin\theta \qquad (5.13)$$

where θ is the angle the pendulum makes with the vertical and mg is the weight of the pendulum of mass m. What makes this different from the oscillatory problem discussed earlier is that the motion is two-dimensional, with components in both the horizontal, x, and the vertical, y, directions. The problem can be written down – in essence we have done it – but it must be taken further to break up the motion into its constituent parts. The outcome would be a set of coupled equations that link the horizontal motion to the vertical.

It is therefore extremely difficult using this approach to calculate the horizontal velocity at any given angle, based as it is on integration of the acceleration. It is possible to look for a set of circumstances under which the motion approximates to something that can be dealt with, as occurs when the angle of the pendulum is small enough to ignore the motion in the vertical direction so that only the horizontal component need be considered. If the length of the pendulum is l, then under these circumstances (corresponding to $\theta \leq 10°$) it is possible to write:

$$F = -mg\frac{x}{l} \qquad (5.14)$$

As m, g, and l are all constants, this equation is analogous to the equation of motion for a mass on a spring described earlier, where instead of the single constant k there are now three. Mathematically this is just a detail that makes no difference to the nature of the solution, which is identical in form. Under the small-angle approximation, as this set of circumstances is known, the solution is simple harmonic motion of the type we have discussed. However, that tells us nothing about the motion at larger angles. Do we need to develop a new theory of mechanics to deal with this problem?

Consider instead an approach based on energy. It is a matter of trigonometry (Figure 5.4) to relate the height lifted to the angle, and hence the potential energy can be calculated:

$$\cos\theta = \frac{l - h}{l} \tag{5.15}$$

or

$$h = l(1 - \cos\theta) \tag{5.16}$$

Using Hamilton's formulation the change in potential energy can be set equal to the increase in kinetic energy after releasing the pendulum, so at any given angle it is possible to calculate the total velocity. As the angle is known, it is again a matter of trigonometry to resolve this into horizontal and vertical components, $v_h = v\cos\theta$ and $v_v = v\sin\theta$ respectively. Hence, the horizontal velocity can be derived as a function of angle. This is not a new theory of mechanics in so far as it does not solve fundamentally different problems from Newton's approach. The problem can be formulated using Newton's laws, it just can't be solved so easily. Under circumstances such as these, finding an alternative approach can be very fruitful.

The *calculational* value of the approach is emphasised above. Sometimes solutions can be obtained in the form of a single equation, as some of the earlier examples have demonstrated; at other times the solution consists of several equations that have to be solved simultaneously. The key to the question, 'New theory or new approach?' lies in our ability to formulate the problem. The formulation of the problem in terms of forces according to Newton's laws might not provide a ready means of computing the motion, but there is essentially nothing wrong with the physics. If necessary, the problem could be solved numerically by dividing time or space into discrete elements and calculating the action of the force across each element. If the element is small enough the force can be assumed to be constant across the element so each calculation then becomes very simple. The only problem,

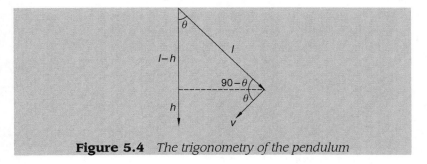

Figure 5.4 *The trigonometry of the pendulum*

of course, is that there will be thousands and thousands of calculations, and a corresponding number of results to record. This sort of operation is best done on a computer, but the fact is it can be done. It is not necessary in this case because an alternative approach based on energy exists. A new theory is therefore not required, but if the results of the calculations, however made, did not match the results of experiments it would be an entirely different matter.

▶ Computation: a new method in physics

The invention of high-power computers has opened up a whole new class of problems to investigation. Computers in themselves do not formulate new theories, but they are important tools in the development of such. The computer can only do as it is programmed, so the essential physics must be formulated. The advantage of the computer is that it allows very large numbers of calculations to be performed in a very short time with the outcome of those calculations being retained. The computer can therefore make accessible calculations that could not otherwise be performed, and though there is perhaps no new physics in the calculations themselves (the new physics, if there is any, will reside in the formulation), the outcome of these calculations may reveal some surprises. It is in the comparison between the calculations and observation that the formulation of new physics arises.

As an undergraduate you are unlikely to be involved with such calculations in any detail, so there is little point in describing them at length. It is sufficient to realise that this constitutes a new approach to physics. The twin pillars of experimental observation and mathematical deduction were developed between the late sixteenth century and late twentieth century, and over that time our views on physics and the way it is done have been modified in the light of experience. The development of supercomputers in the 1950s and 1960s heralded the onset of this latest change, and the explosion in computing power that took place at the end of the twentieth century and still continues today, has consolidated the change.

What these machines allow us to do is to subdivide space and time into small elements, so that the detailed transport of matter or energy over distances can be examined. Thus we can simulate weather patterns, the expansion of matter in the big bang, nuclear explosions, heat diffusion during laser treatment, the properties of crystals, and so on. We can calculate these properties with a resolution hitherto unattainable because the interactions between particles, or of energy with matter, can be followed in great detail. It is not necessarily that we do not understand the physics in these cases – as discussed, these interactions have to be programmed – but that we are unable to calculate these effects in such detail without the aid of a computer. Where such calculations give us an insight that did not previously exist we have a new way of looking at the world.

Example 5.1 The quantum mechanics timeline

In order to put these ideas about the practice of physics and the development of theories into context, consider the quantum mechanics timeline. Much of the work on the discovery of nuclear and sub-nuclear particles has been omitted, not because it has not contributed to the development of quantum mechanics, but because it gives a slant to quantum mechanics that emphasises these aspects at the expense of others. Instead, developments in solid-state physics not normally included in the timeline have been mentioned because they illustrate not only the emergence of these ideas out of the application of quantum mechanical concepts, but also the development of a separate and distinct field that fully depends on quantum mechanics for its existence but is as yet not part of the mainstream of modern quantum theory.

The timeline may well start even before the turn of the century, with the recognition that classical theories were unable to predict the blackbody spectrum, which is an example of Popper's falsification, but we start instead on a constructive note, at the development of the theory that did predict the blackbody spectrum.

1900 Max Planck suggests that radiation is quantized, based on his calculations of the properties of black bodies.

1905 Albert Einstein proposes a quantum of light which behaves like a particle in order to explain the photo-electric effect.

1913 Niels Bohr succeeds in constructing a theory of atomic structure based on quantum ideas. The idea was not fully self-consistent in that no real theoretical basis for the quantum picture was proposed.

1916 Arnold Sommerfeld introduces the magnetic quantum number, and four years later the spin quantum number was introduced, which led to the discovery of electron spin.

1923 Arthur Compton discovers the quantum (particle) nature of x-rays, thus confirming photons as particles.

1924 Louis de Broglie proposes that matter has the properties of a wave.

1925   Wolfgang Pauli formulates the exclusion principle for electrons in an atom.

1926   Erwin Schrödinger develops wave mechanics. This is the first real theoretical explanation of why matter and energy appear to be quantized. Max Born gives a probability interpretation of quantum mechanics, called the Copenhagen Interpretation. The name 'photon' is proposed by G. N. Lewis for a light quantum.

1927   Werner Heisenberg formulates the Uncertainty Principle. Davisson and Germer demonstrate wave–particle duality directly by diffracting electrons from a nickel crystal.

1928   Paul Dirac combines quantum mechanics and special relativity to describe the electron. Sommerfeld models the electronic properties of metals by treating the electron as a quantum particle obeying Fermi–Dirac statistics. The model is so successful that many concepts are still used today in semiconductor physics. Bloch showed that a perfectly periodic structure allows quantum particles such as electrons to pass unimpeded, and hence the source of resistance in electronic solids is caused by the natural deviations from periodicity due to imperfections.

1931   Paul Dirac realised that the positively-charged particles required by his equation are new objects. Naming them 'positrons', they have the same properties as electrons but with a positive charge. This is the first example of an antiparticle.

1947   Procedures are developed to calculate the electromagnetic properties of electrons, positrons and photons. Feynman diagrams are introduced. The transistor is invented by Bardeen Brattain and Shockley.

1957   Bardeen, Cooper and Schrieffer develop a theory of superconductivity.

1958   Leo Esaki observes quantum mechanical tunnelling in semiconductor structures, for which he received the Nobel prize in 1973. This work opened up the field of semiconductor engineering and the subsequent development of quantum technology based in atomically abrupt interfaces in semiconductor materials.

1960   The ruby laser was developed by Maiman.

1962   The first semiconductor laser was demonstrated.

The stories behind these headlines are much more interesting than the bald statements suggest. Planck was at first reluctant to believe the implications of his formulations because they were so at odds with the accepted picture of the physical world. However, calculation of the blackbody spectrum had so far eluded those using classical concepts, and Planck was forced to accept the inevitable, an example perhaps of *deduction* and a demonstration of the compelling logic of mathematics when backed up by experiment.

Bohr's model of the atom is an example perhaps of induction, and illustrates the weakness of the inductive model in physics. Experiment had clearly shown that light emitted from atoms has definite frequencies, and it was possible to derive simple mathematical relationships between these frequencies. As an observation this could be taken as fact, but the fact itself is interesting only because, first, it showed that the classical electrodynamic model of the electron in an atom could not possibly hold true, and second, because it provides the data that a better theory should reproduce. Bohr's model reproduced the data by assuming that electronic orbits were quantized but without providing any explanation as to why it should be so. This is an example of a *phenomenological* model, because it is based on the experience of the phenomenon. It contains an *implication*; that electronic orbits are quantized, but that implication would remain no more than an interesting possibility unless a more fundamental model which explained in some measure the reason for this quantization.

Such a model came about through the development of the Schrödinger equation, which took account of the wave–particle duality. This had been first proposed in Einstein's model of the photoelectric effect, then by De Broglie, and demonstrated by Davisson and Germer, which demonstrates the predictive power of a theory and the ability of experiment to confirm it. Thus matter could be treated as waves, and Schrödinger's idea was that electron waves in an atom would destructively interfere with themselves if the orbit was not a whole number of wavelengths. This constitutes the first attempt to *explain* quantization.

We then see the application of this theory to the solid state, a field largely ignored by many working in quantum mechanics at the time. Solidstate theory developed rapidly and even began to contribute to quantum mechanics via the observation of tunnelling, and later the development of quantum structures through atomic scale precision in the deposition and fabrication of semiconductor structures. Such a development was called by Esaki, in a paper of the same title, 'do it yourself quantum mechanics', and freed the investigator from the restrictions imposed by nature.[8] It became possible to specify the details of a quantum system in order to investigate particular phenomena and build it within the solid state. The computational details of such structures are very complicated and this is an example of the contribution that high-power computers make not only to physics but also to technology. High-speed transistors, diode lasers, photo-detectors, and so on are all examples of quantum devices, and the very technology on which this book was written could not have been developed without computational solid-state physics based on quantum mechanics.

▶ Summary

The purpose of this chapter has been to identify a working model of physics that will enable you as a student to see:

- The relationship, as well as the difference, between maths and physics.
- The importance of mathematics as the symbolic logic in which arguments are constructed.
- The compelling logic of mathematical deduction when backed up by experimental evidence.
- The hidden assumptions in an argument.
- The hidden implications of an argument.
- The necessity for open-mindedness and the value in taking a different approach to a problem.

Along the way several ideas have been developed, such as the existence of distinct fields linked by common themes (hierarchies) underpinned by a theoretical framework (paradigm). Not everyone will agree with these ideas, and ultimately you may develop different views. The thoughts presented here are therefore not intended to be rigorous philosophical arguments, but merely to suggest a way of looking at physics that is useful to help you as a student to come to a deeper understanding of what it is you are doing and trying to achieve.

▶ Notes

1 Andrew Elby, *American Journal of Physics*, vol. 67, 1999, pp. S52–S57 (emphasis added).
2 *Ibid.*
3 Eunsook Kim and Sug-Jae Pak, *American Journal of Physics*, vol. 70, 2002, pp. 759–65.

4 Daniel F. Styer, Miranda S. Balkin, Kathryn M. Becker, Matthew R. Burns, Christopher E. Dudley, Scott T. Forth, Jeremy S. Gaumer, Mark A. Kramer, David C. Oertel, Leonard H. Park, Marie T. Rinkoski, Clait T. Smith and Timothy D. Wotherspoon, *American Journal of Physics*, vol. 70, 2002, pp. 288–97.

5 Stillman Drake, *Galileo*, Oxford University Press, 1980.

6 Elby (1999), *op. cit.*

7 M. Boas, *The Scientific Renaissance 1450–1630*, London, Collins, 1962.

8 L. Esaki, *Physica Scripta*, vol. T42, 1992, pp. 102–9.

6 Mathematical Modelling

Simple harmonic motion has been used as an example so far of a mathematical model, but the range of mathematical techniques required by the physicist is far greater than is represented by this single example, or indeed that can be included in this chapter. Instead, the purpose here is to present further formulations of basic mathematical models of physical phenomena and in some cases to solve them. In this way some of the most basic mathematical techniques can be seen in application, and, furthermore, the interconnectedness of ideas in physics can be demonstrated.

A mathematical model requires the three components described in Chapter 5:

- Formulation.
- Mathematical development of the solution.
- Interpretation of the solution.

Formulation and interpretation are the key aspects emphasised once again. It will be shown that the differential equation, that is, an equation containing one or more terms expressed as a differential, is one of the most common methods employed by physicists to formulate a problem. The differential equation is arrived at by setting down the problem in its most basic form, for example equating forces in Newton's second law of motion. There are a range of techniques that can be used to solve these differential equations, by which is meant the derivation of an equation or set of equations that together both satisfy the formulation equation and describe the quantities of interest. In the case of SHM, as we have seen, the mathematical *formulation* of the problem is:

$$\frac{d^2y}{dt^2} = -\omega^2 y \tag{6.1}$$

The *solution*, or at least *a* solution because there is more than one, is:

$$y = A\sin(\omega t) \tag{6.2}$$

which describes the motion y, the subject of the differential equation (6.1).

▶ Solving differential equations

This is a subject that requires a textbook in itself. In essence, the equation has to be integrated but there are many equations that do not submit to simple integration. There are some that have recognisable solutions, such as Bessel's equation for example. Bessel's equation is a particular form of a differential equation that has well-defined solutions called, not surprisingly, Bessel functions. If a problem can be formulated in terms of a well-known equation for which the solution already exists, your workload is considerably reduced. All you have to do is recognise it. Mathematical tricks such as this therefore require familiarity.

As an undergraduate you are very unlikely to encounter such complicated equations until your third or fourth year, and of course you will not be expected to develop the mathematical skills on your own. The same applies to other mathematical techniques, such as *linear algebra*. Linear algebra is the algebra of matrices, which are very commonly used to express the solutions of differential equations. Matrices are very useful, but their use is not easy to describe in a few words. As with most mathematical techniques, the subject has to be developed through very simple examples before the application to real physical problems can be attempted. In short, there is no room to present such techniques here. The emphasis is very much on you to study and learn such techniques as are required of you as an undergraduate.

Studying mathematics

Techniques such as integration and linear algebra belong to the realm of mathematics rather than physics. Quite likely they will be taught as mathematical techniques in their own right, to be learnt as such, and with their application to physics problems being given much later. As emphasised throughout, the approach to mathematics requires practise, practise, and more practice. Mathematics is a branch of logic. It has to be approached as such, and not with a view to seeing the physical significance of mathematical steps. This is the natural approach of the physicist, but it doesn't always serve with mathematics.

▶ Mathematical software

The late twentieth century saw the development of a number of complicated mathematical software packages and languages such as MathCad, Matlab, Maple, and Mathematica. These software packages do not require programming in the sense of writing a code in FORTRAN or C++. Of course they have their protocols that have to be learnt, but they are intended to allow the specification of a mathematical problem in relatively simple terms. In MathCad, for example, the input is in the form of a text editor and the problem is written using conventional mathematical symbols, just as you would write it in your own notebook. Mathematica also uses symbolic notation and moreover, can select an algorithm to solve a particular problem.

These programmes are designed for use by professionals working in technical fields like engineering or physics. They will handle complicated integrations, matrix multiplication, and a host of other mathematical operations. As an undergraduate you could certainly use them to solve most of the mathematical problems that you will face. In addition they provide a valuable resource for learning and understanding mathematics, because they provide a ready means of performing difficult operations quickly. However, a word of caution is required. It is one thing to use such programmes in a professional capacity when you have already developed an awareness of mathematics and physics, it is quite another to use them as an alternative to acquiring mathematical skills. *The process of formulating a problem and interpreting the results requires an understanding not only of physics but also the relationship between the maths and the physics.* There can be no substitute for this type of understanding, whatever may happen to the mathematics in between.

Computing a solution

Mathematical software is especially useful for computing the solution of a model. Conventionally an equation, or set of equations, would have been computed after writing a dedicated computer programme, an approach still favoured by some. Personally, I question whether this is the most useful method for undergraduates. Nonetheless, computation in the form of programming has formed, and no doubt will continue to form, a significant part of an undergraduate physics education. Programming has its uses, no doubt about it, but it is simpler and more effective to use mathematical software to compute the solutions to most problems you will face as an undergraduate.

Formulating a model

The analytical/deductive modelling described by Descartes (Chapter 5) is of most interest in physics. Formulating such a mathematical model requires first and foremost that the assumptions are identified. As described in

Chapter 5, there can be hidden assumptions that might not be explicitly stated at the outset and therefore not expressed mathematically. These assumptions usually assume an importance when the model fails to describe the physical phenomenon being modelled.

There are some simple rules that can be employed:

- Start simple and work up. Make simplifying assumptions if possible, so that at least you have a chance of finding a solution, albeit a simple one, before making a model too complicated.
- Express the physical reality in mathematical form.
- Stick to the generalities by using symbolic notation wherever you can. Leave the specifics where possible to a consideration of a specific solution.
- Identify boundary conditions, that is particular values of certain variables at specific locations and at specific times, at the outset. This helps to define the problem in many cases.

▶ Types of models

Although mathematical/deductive models are of most use, completeness requires at least that inductive/empirical models be considered. Experimental, or empirical, models take the results of known experiments and attempt to express the results in a mathematical form. The models are usually descriptive rather than explanatory in so far as they provide a method for understanding what will happen rather than why it will happen. It's a form of inductivism; inducing the general from the specific, but as described in the previous chapter this sort of reasoning has its limitations. Not least, experimental models are not especially good at revealing the circumstances under which deviations from the observed behaviour will occur. For this we have to have a deeper understanding of the phenomena, and this is the purpose of the theoretical models.

Theoretical models are deductive and as such are inherently mathematical, though of course the complexity of the mathematics can vary extensively from one model to the next. Theoretical models are intended to be explanatory. For example, an electron can be described mathematically by means of quantum mechanics, put in a particular environment such as a solid, and its behaviour predicted under certain conditions. This would be considered explanatory to a great extent, as there is an answer to many 'why?' questions that may be asked about solids; why do some conduct, why are some insulators, why are some highly reflecting while others are transparent? All these follow from the properties of electrons in solids, so a model that correctly describes their behaviour 'explains' many other phenomena.

This brings to the fore once again the hierarchical nature of theories, as discussed in the previous chapter. If further questions are asked about the nature of the electron we would find that we are unable to answer them fully. At some level the 'why' is not complete, but within the context of solid-state physics it is not important. The exact nature of an electron has no bearing on the theories of solid-state physics. We simply take as granted that an electron exists. We describe it mathematically and this serves as the basis for the explanation of phenomena of interest. The exact nature of an electron is obviously a question of great interest to many physicists, mostly those working in the field of particle physics, but even if the question were answered to the satisfaction of all it would not alter the theories of solid-state physics.

Mathematical descriptions thus do not extend to the whole of reality. There must be a starting point, which is of course the statement of the basic assumptions of the model. These assumptions may be logical truths, as revealed by mathematics, or they may be experimental facts. Hence, in solid-state physics we take it as fact that an electron exists; it has a mass and a charge and it causes a deflection on a meter. A model based on our experience, by which we mean experiment, is said to be *phenomenological*, but there are degrees of phenomenology. Sometimes experimental facts (phenomena) form the basis of an explanation which is then deductive. In other cases experimental facts have to be inserted into an otherwise explanatory model, leading to a mixed phenomenological/deductive model.

Empirical models

An experiment provides us with the 'hard' fact of our experience, but it must do so in such a way that the subjectivity that accompanies all our observations is minimised. By this I mean simply that if you and I measure the same phenomenon our results may differ slightly, and so all measurements have associated with them a degree of uncertainty that is essentially an expression of the subjectivity (see Chapter 3). Granted that experiments can be performed, measurements made, and conclusions drawn to specified accuracies, what can those measurements tell us about the world in which we live?

Very much depends on the reason for doing the experiment in the first instance. An experiment may be performed in support of a theory, in which case some deduction will already have occurred; the model has already been built and the experiment is simply designed to test it. What about the situation where an experiment is simply performed to find out some previously unknown information, much in the manner of Galileo's experiments? We will have a description of nature based on the evidence of our senses. Such a description is called an empirical model.

In order to illustrate the idea, consider Ohm's law as an example. This is not meant to be a historical account of how this law was developed, but to demonstrate a point about this law. Suppose we know nothing about resistance, but have a number of wires, a voltage source, and a current meter. We could apply a voltage to one of the wires and measure the current flowing. If the voltage is varied different currents would be measured, with the current increasing in direct proportion to the increase in voltage:

$$i \propto V \tag{6.3}$$

There must be a constant of proportionality which we now know is called the conductance G, its inverse being the resistance R.

It is impossible to know how general this observation is. We could guess that other wires might behave in like manner, but without doing the measurements we would be left in the position of wondering whether, however improbably, there was something unique about the wire we had chosen. If we did the same to other metal wires it might be able to deduce something general, but if the wires were randomly constructed of different metals, lengths, and diameters, it would be very difficult to determine anything of substance. There would be nothing systematic in the measurements and hence nothing systematic in the results. However, using wire of one particular metal cut from a particular reel so that all the wires were of an average diameter, a systematic study would reveal that for a given voltage applied, the current flowing would decrease with increasing length, l. If the diameter of the wire was then varied systematically we would find that resistance decreases in inverse relation to the cross-sectional area A.

At this point we would have enough data for our first simple model:

$$R \propto \frac{l}{A} \tag{6.4}$$

or

$$R = \frac{\rho l}{A} \tag{6.5}$$

where ρ is a constant of proportionality called the *resistivity*. In this form the model appears about as fundamental as it can be and the resistivity appears to be a property of the metal that is independent of its dimensions. We would be tempted to believe that all metals have a resistivity, so there is an element of induction. The model doesn't tell us much about other metals, though, so it is not predictive in this sense. Each metal has to be characterised in its own right, but having characterised a metal in this manner it would be possible to predict the current flowing in any piece of wire. This is as useful as the model gets, though. It is purely descriptive. We have concluded something about the physical world but it is not an explanation for the phenomenon.

Phenomenological models

Phenomenological models are those that are based primarily on experience, but they are distinguished from the purely descriptive empirical models by a deductive element. Thus, phenomenological models are often predictive in some sense so they allow a development of our understanding through subsequent experimentation, but they are not really explanatory. By way of example, let's continue with the theme of resistance, from which it is possible to develop some ideas about the nature of conduction in metals.

The simple definition of a current is that it is the amount of charge flowing per second, so let's construct the most basic model of current flow based on electron flux. This is a standard argument in transport problems. Consider a length of wire l of cross sectional area A, with a density of electrons n (Figure 6.1). The total *number* of electrons in the wire is $n_l = n \cdot l \cdot A$. If these electrons are travelling with an *average* velocity v in one direction along the wire, the time taken for an electron to traverse the length of the wire is $t = l/v$. Hence in one second, the number n_s of electrons passing through the end of the wire, assuming a ready supply, is:

$$n_s = \frac{n_l}{t} = \frac{n \cdot l \cdot A}{l/v} = n \cdot A \cdot v \tag{6.6}$$

The total current is just the product of this number and the electronic charge q (sometimes written as e); that is:

$$i = v \cdot A \cdot n \cdot q \tag{6.7}$$

It only remains therefore to determine the average velocity of the electron and we can begin to understand something about current.

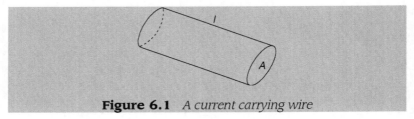

Figure 6.1 *A current carrying wire*

If a charge experiences an electric field, it is subjected to a force that must, by Newton's second law, lead to an acceleration. Assuming the electric field to be nothing more than the ratio of applied voltage to the length of the wire, the force is:

$$F = qE = \frac{qV}{l} \tag{6.8}$$

If the electron is accelerated continuously from one end of the wire ($x = 0$) to the other ($x = l$) then the velocity at the end of the wire, v_l, will be, from Newton's kinematic equation,

$$v_l^2 = u^2 + 2\frac{qV}{m_e l} \cdot l = \frac{2qV}{m_e} \tag{6.9}$$

where $u = 0$ is the initial velocity at $x = 0$, V is the applied voltage, and m_e is the electron mass. The average velocity will simply be $v_l/2$, so the current will be proportional to the *square root* of the applied voltage. This is not Ohm's law, so the charge cannot be accelerated from one end of the wire to the other.

If it is not accelerated from one end to the other then something must be acting to resist the flow of electrons through the metal (hence the term resistance). We can postulate some sort of collision mechanism as being responsible. This is the most simple mechanism consistent with the facts. As the electrons collide they lose energy, so the electrons have an average velocity that must depend only on the strength of the electric field. Therefore the following argument can be constructed using the concepts of classical mechanics:

- An electron exists.
- The electron behaves like a particle.
- The electron accelerates in response to an electric field.
- Assume at some point the electron loses its energy through an unspecified collision.
- An average (mean) time to collision leads to an average velocity for the electrons.
- If the mean time to collision is independent of the velocity, that is, it is a property of the metal, the velocity of the electrons will be proportional to the electric field.
- This constant of proportionality can be called the mobility.

This argument is still not explanatory in so far as the collision mechanism is not specified. It doesn't tell us, therefore, anything about the nature of the resistance but uses it as an experimental fact together with the existence of an electron as a particle to assert a collision process during conduction. It is therefore a phenomenological model.

Although the model is not explanatory, it is predictive. It is an attempt to provide a framework for understanding some aspects of conduction and in some respects is remarkably successful. It is possible, for example, to show that the ratio of thermal conductivity to electrical conductivity has a certain temperature dependence, a prediction of the model born out by experimental observation. In other respects it is deficient, and to overcome these deficiencies it is necessary to go to another level of modelling.

Ab initio **models**

The phrase *'ab initio'* is Latin and means 'from the beginning'. An *ab initio* model is one in which an explanation is framed in terms of the most basic ideas. The resistance of a solid can be understood in terms of lattice imperfections, and the potential 'seen' by an electron in a lattice varies according to the position of the electron with respect to the atoms. If the atoms are effectively positively charged, as they will be in a metal because the electrons that take part in the conduction process have been 'lost' by their parent atoms, the lowest potential will coincide with the position of an atom and the highest potential will exist between atoms (Figure 6.2). In other solids the situation may be different, but if the solid is crystalline we would expect the potential to be periodic.

It turns out that if we consider the quantum mechanics of an electron wave propagating through a *perfectly* periodic potential represented by the regular array of atoms, the electron is not scattered. Resistance therefore arises from imperfections in the lattice. This is an *ab initio* model of resistance because it starts from the very basic fact of an electron in a periodic potential. There are no assumptions other than the fact that an electron exists, which, as already argued, is a phenomenological aspect of

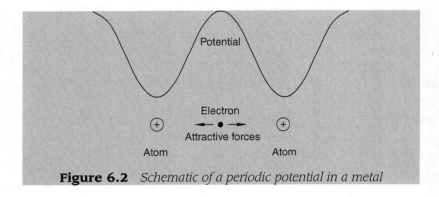

Figure 6.2 *Schematic of a periodic potential in a metal*

the model. However, the phenomenology is at the lowest possible level so it does not alter the explanatory nature of the theory.

From the general to the specific: modelling systems

One of the features of *ab initio* models is that they are very general, but their very generality means that they cannot easily be applied to specific cases without some modification. The model of resistance described above is not a specific model that allows us to calculate the resistance of any particular solid; rather it is a general physical principle that tells us the origin of resistance. In order to calculate the resistance of a given metal we would have to know the defect structure of the metal; for example the density and distribution of impurities, and the density of missing atoms, or vacancies. Some of these quantities can be calculated using thermodynamic arguments, others might have to be introduced phenomenologically. Nonetheless, we would still regard it as an *ab initio* model of resistance because it starts from the most basic formulation possible; the quantum mechanical description of an electron propagating in an imperfect lattice defined by thermodynamic data.

▶ Mathematics and modelling

It is quite apparent from the preceding discussion that the complexity of the mathematics required for different models can vary widely. As mathematical techniques have developed, so too has our understanding of the physical world, *because our physical world is described in mathematical terms*. The mathematics of Ohm's law, for example, is pretty basic, but as described in Chapter 5, at the time of the Renaissance even this mathematics would have been difficult. In consequence, whilst we would regard it as true that empirical models are of limited use, we would be hard pressed to say that this has always been the case. It is easy to imagine that at the time when Ohm's law and Newton's laws of motion and gravitation were formulated they would have constituted a revelation in the understanding of the physical world.

The key to mathematical modelling, then, is not only a sure grasp of mathematical techniques, but also a sure grasp of the theories and concepts of physics. The two go hand-in-hand. The following pages contain some mathematical models of basic physics concepts. As models go they are mathematically quite simple, but they find widespread application in physics, and introduce some mathematical concepts that may be unfamiliar to you. The intention therefore is two-fold; to introduce these mathematical ideas by way of example and also to illustrate the extent to which these ideas permeate the whole of physics. We start with a first-order differential equation, exemplified by the physics of radioactive decay.

▶ Radioactive decay

Radioactive decay is a classic example of a first-order differential equation. It is possible to show that after any interval of time t the number of radioactive atoms N in a sample is given by:

$$N = N_0 e^{-\lambda t} \tag{6.10}$$

where N_0 is the number of atoms at time $t = 0$, that is when the clock started, and λ is some constant. In order to understand this equation the following two statements are asserted:

- in any radioactive sample the rate of decay depends on the number of atoms present; and
- the time taken for any given fraction of the atoms to decay is independent of the size of the sample and depends only on the material.

These statements are not immediately obvious. In fact they may even appear to be wrong. Does the statement that the rate of decay depends on the number of atoms present imply some sort of interaction between the atoms, so that one atom 'knows' that another is present, or does it simply mean that the more atoms there are to start with the more will decay in a given time? It all depends on what is meant by the rate of decay. The second statement is a logical consequence of the first.

The statistics of radioactive decay

Let us suppose that we could extract a number of atoms, say 100, from a sample and line them up in such a way that they could be observed and counted. We know nothing about the microscopic details of radioactive decay, so there is no way of predicting when a particular atom will decay. It is, as far as we know, a purely random process. That does not mean to say that there is no mechanism; there is, but we don't know anything about it. Let's stop the clock after half of the atoms have decayed, so we are left with a sample now containing 50 atoms. Our second statement above says that it will take just as long until 25 atoms decay, and then the same length of time for the next 12 (strictly 12.5 but we can't have half a decay). Does this mean the process is slowing down?

Suppose we had another, identical sample of one hundred atoms and watched this decay. The time may well be slightly different because of the probabilistic nature of the process, but there is no physical reason to suppose that one sample will behave fundamentally differently from another. With 100 such samples there would be enough data to take an average and evaluate the error on this average using the Poisson distribution. It has already been shown that averaging over 100 samples of 50 counts is identical to counting once for 5,000 counts. The mean count rate and its error are exactly the same. Statistical fluctuations aside, the time taken for one sample to decay by half is the same as that taken by another, and another and so on, independent of the size of the sample. Looking at the totality of events, the time taken for a giant sample of 10,000 atoms to decay by half is the same as the time taken for each individual sample of 100 atoms to decay by half. This is called the *half-life*.

Independent events
The total number of events in any given time interval simply depends on the number of atoms present in the first instance. It would of course be different if one atom influenced another, as happens with uranium for example. If sufficient uranium exists together in pure form then one decay can cause another, and a self-sustaining chain reaction occurs. This is the so-called critical mass, but the conditions have to exist for the by-products of the reaction, that is the neutrons, to interact with other uranium atoms and stimulate a fission reaction. If these conditions do not exist each atom essentially decays independently of every other. These conditions obtain in most radioactive decays, so it follows that the total number of events in a given time is simply a reflection of the total number of atoms present in the first instance. It also follows that the history of the sample is unimportant and that any sample of a radioactive element will behave like any other. The two initial statements that seem counter-intuitive at first are in fact very reasonable.

Differential equations
So far this is a purely logical argument. It contains a basic assumption that may or may not correspond to the reality; namely, the individual decay events are independent. The next step in the process is to formulate this mathematically in order to arrive at a testable hypothesis. If experiment did not match the hypothesis it would then be necessary to revisit the assumptions, but that is another matter. To continue, the rate of decay, that is the number of atoms that decay in a given small interval of time, is proportional to the number of atoms present, so:

$$\frac{dN}{dt} \propto N = -\lambda N \tag{6.11}$$

where λ is the constant of proportionality and is negative because the rate of decay is decreasing. This is a first-order differential equation, so-called because only the first differential is involved. It is one of the simplest differential equations, and integration is straightforward. Rearranging,

$$\frac{dN}{N} = -\lambda dt \tag{6.12}$$

from which it follows that:

$$\int_{N_0}^{N} \frac{1}{N} dN = -\lambda \int_{t=0}^{t} dt \tag{6.13}$$

The integral of $1/N$ is well-known and has the form $\ln(N)$, so:

$$[\ln(N) - \ln(N_0)] = \ln\left(\frac{N}{N_0}\right) = -\lambda t \tag{6.14}$$

Taking exponentials leads to the starting point of this discussion.

Exponential decays of this sort are quite common in physics. Wherever microscopic events are independent from each other, so that the totality is simply the sum of all the individual events, an exponential decay will occur. If the events are correlated, so that one depends on another in some way, the mathematics must necessarily be different and so too must the solution.

▶ Simple harmonic motion

Simple harmonic motion (SHM) has already been described in relation to the formulation and interpretation of mathematical theories. Here the intention is to deal with the mathematics of SHM itself, which is a concept not related to any specific physical reality. In short, it is a mathematical abstraction.

Motion implies velocity and acceleration, and in the physical world the latter implies a force. It is possible therefore to define mathematically the type of force that *must* lead to SHM by analysing the details of the motion. This is the abstraction. Although talk of forces and acceleration implies real objects, the motion itself, including the acceleration, is defined without reference to a physical object. Translation of a force into a real acceleration requires a mass through Newton's second law, but there is no reason to suppose that just because we have mathematically defined the nature of the force required for SHM such forces exist in nature. SHM is therefore a mathematical abstraction. The physics lies in finding those situations in which the mathematics applies.

The mathematics of SHM

In order to derive the force leading to SHM, let's start with the classic text book definition. Suppose we have a vector of length A rotating in an anti-clockwise manner, as shown in Figure 6.3. A vector has both magnitude *and* direction so the term describes very well the mathematical situation. In this case the magnitude is not changing but the direction is, so there must be an acceleration. The vector will describe a circle of radius A and makes an angle θ with the horizontal, where of course θ changes with time. The projection of A on the horizontal and vertical respectively is given by the simple trigonometric relations $x = A \cdot \cos\theta$ and $y = A \cdot \sin\theta$. These two are 90° out of phase with each other, so that as x decreases y increases. If θ changes uniformly with time, that is:

$$\theta \propto t = \omega t \tag{6.15}$$

then by definition,

$$y = A\sin(\omega t) \tag{6.16}$$

This is simple harmonic motion. If $\theta = 0$ at $t = 0$, then $y(0) = 0$, rising to A when $\theta = 90°$, returning to 0 again for $\theta = 180°$, falling to $-A$ at 270°, and back again to zero. The vector A will make one complete rotation in $2\pi/\omega$ seconds. ω is known as the angular velocity and is simply the rate at which the angle changes. As described, there is no reference to the natural physical world here. Of course, it would be possible to construct a machine that did something very similar, but that would be wholly artificial. You may ask why the transverse motion,

$$x = A\cos\theta \tag{6.17}$$

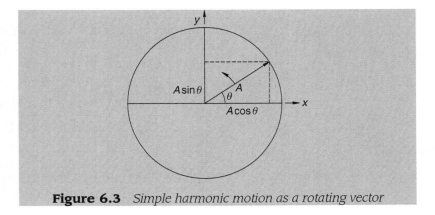

Figure 6.3 *Simple harmonic motion as a rotating vector*

was not chosen. Physically there is a good reason to choose (6.17) rather than (6.16), but mathematically it makes no difference apart from the starting point, so let's continue with the motion in y.

Given that y varies with t, that is, we have a function $y(t)$, it is possible to define a velocity dy/dt and an acceleration d^2y/dt^2. Proceeding straight to the latter,

$$\frac{d^2y}{dt^2} = -\omega^2 A \sin(\omega t) = -\omega^2 y \qquad (6.18)$$

This is a *second-order linear differential equation* and is pretty much all we need to know. Acceleration is proportional to the force applied, so

$$\frac{d^2y}{dt^2} \propto F \qquad (6.19)$$

and

$$F \propto -y \qquad (6.20)$$

The force increases in magnitude as y increases, but the negative sign indicates that the force is acting in the opposite direction to the motion (Figure 6.4). The mass and ω^2 omitted from (6.20) are not important to the central argument; they simply affect the magnitude of the force but not the fact that the force always acts in the opposite direction to the motion and increases in magnitude the further away the mass moves from the centre. Such a force is called a linear, central force, because it varies linearly with distance from the centre and acts towards the centre. Any physical system in which such a force appears will undergo SHM if disturbed from its equilibrium position, that is from where the net force is zero.

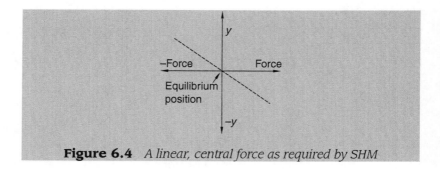

Figure 6.4 *A linear, central force as required by SHM*

▶ From mathematics to physics

It is worth spending a moment to contemplate what has just been described here. The sine function is a mathematical consequence of the rotating vector, not of the force. Physics often proceeds by the opposite route; by establishing the equations of motion and deriving the corresponding trajectory, but here we have started with the simple notion of the one-dimensional projection of a two-dimensional motion. This is mathematically exact and rigorous, but done without reference to any physical reality. There is no reason why physical systems should behave in this manner but, remarkably, many physical systems do. A mass on a spring, a pendulum (if the angle is small), a molecule; the list of physical systems characterised by a linear, central force is extensive.

It is interesting to compare SHM with radioactive decay. Obviously the mathematics is different, but there is also an important difference in the underlying relationship between the maths and the physics. Both are examples of the mathematical deductive process described by Descartes, in which the logic of mathematics is used to find a description of the physical world. However, it is possible to test for SHM simply by looking at the nature of the force because of the exactness, and the abstractness, of the mathematics. In other words, there is a precondition of SHM that can be tested. The precondition for the model of radioactive decay is the independence of the atoms, which cannot be tested directly. It is impossible to line up atoms and watch them, so the only thing that can be done is to measure the count rate, and by comparing the observed behaviour with the predicted, *infer* the validity of the assumptions. In some cases it will not even be possible to measure the half-life directly, as it is too long. Some elements, such as uranium, have a half-life of the order of 10^9 years, and these have to be extrapolated from the very slow rates of decay measured in the laboratory.

This may seem a small difference, but it is an important difference nonetheless because the manner in which we approach experimental measurements of the phenomena is determined by it. It is because of such differences that it becomes very difficult to identify a *single* process for physics. It would be very nice to be able to say that physics is done this way or that, but in truth it is done in many ways. The only principle that can safely be offered is to understand in detail the specific relationship between the mathematics and the physics, the hidden assumptions and implications, and to proceed on a case-by-case basis.

▶ Linearisation

The exact mathematical condition for SHM is not always found in mechanical systems, yet it can be demonstrated mathematically that *any* bound system will undergo SHM under the right conditions. Is this a contradiction? No, it is an example of *linearisation*; the process of reducing a complex non-linear problem to a simple linear problem susceptible to mathematical analysis.

As before, there is more than one way to approach the problem. A specific non-linear force can be examined to see if the circumstances under which it becomes linear can be found, or the problem can be approached using the energy of the system to define a set of conditions – quite general conditions in fact – that must apply for a motion to appear as SHM.

Linearisation of a force: the pendulum

The simplest physical system to examine is that of a pendulum. It is fairly easy to show that if the pendulum makes an angle φ with the vertical, the force directing the motion toward the centre, that is the bottom of the swing, is proportional to $\sin \varphi$. This is no longer a linear force. However, if the units of the angle are expressed in radians then for very small angles $\sin \varphi$ approximates to φ. Radians are in fact a more natural unit of angle than the degree. The idea that there should be 360° in a circle has come down from Babylonian times where it was believed that there were 360 days in a year and the number 360 appeared in so many different aspects of the civilisation's culture. It is therefore a somewhat artificial notion to have 360° in a circle.

The radian is more natural because it is defined using a basic property of the circle, namely that the number of radii that fit exactly into the circumference is 2π, so if the circle is divided into 2π equal segments, the length of arc in each segment will equal the radius. This defines the radian; it is the angle in each of these segments, and is equivalent to 57.3°. There are thus 2π radians in a complete circle, and the angle of any segment can be calculated simply by taking the ratio of the length of arc to the radius.

In relation to the pendulum, then, the length of arc l through which the pendulum swings is simply the product of the radius and the angle (Figure 6.5). The motion is clearly two-dimensional so in describing the acceleration it is necessary to consider the rate of change of l rather than x or y. The force acting to restore the pendulum to the centre is of course gravity, but strictly gravity acts vertically so the force must be resolved in the direction of motion, that is at a tangent to the radius r. It is just trigonometry

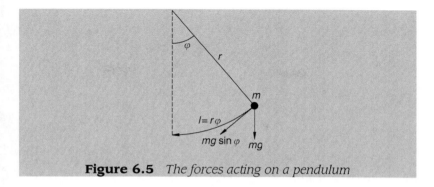

Figure 6.5   *The forces acting on a pendulum*

to show that this is $mg. \sin\varphi$. Therefore, using Newton's second law,

$$m\frac{d^2l}{dt^2} = -mg \sin\varphi \qquad (6.21)$$

where, as before, the force is negative because it acts towards the centre.

  This is a second-order differential equation, but it is clearly not linear and so the solution cannot be SHM. An equation such as this is in fact quite difficult to solve analytically. Numerically it can be solved easily enough, as described in Chapter 5, but how would this problem have been tackled in the pre-computer age? There are two possible answers. The first is to investigate the mathematics to see if there are methods that can be developed to help solve the problem. The second is to linearise it. As you might have guessed this involves making the angle φ small. It is a mathematical property of the sine and cosine functions when the angle is expressed in radians that they can be expanded as series of higher-order terms. Thus:

$$\sin\varphi = \varphi - \frac{\varphi^3}{3!} + \frac{\varphi^5}{5!} - \frac{\varphi^7}{7!}\ldots\ldots \qquad (6.22)$$

and

$$\cos\varphi = 1 - \frac{\varphi^2}{2!} + \frac{\varphi^4}{4!} - \frac{\varphi^6}{6!}\ldots\ldots \qquad (6.23)$$

which means that for small φ:

$$\sin\varphi \approx \varphi \qquad (6.24)$$

The restoring force now becomes:

$$mg \cdot \sin\varphi = mg \cdot \varphi = \frac{mgl}{r} \qquad (6.25)$$

and the differential equation is *linearised* to become:

$$\frac{d^2l}{dt^2} = -\frac{gl}{r} \qquad (6.26)$$

which is directly comparable with the equation for SHM.

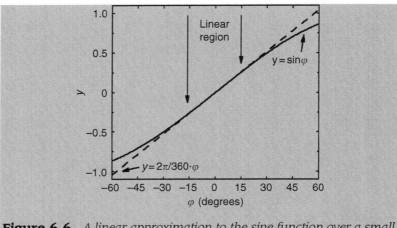

Figure 6.6 *A linear approximation to the sine function over a small range of angles*

The range of φ for which this approximation is valid is surprisingly large. The principle condition is that φ must be considerably less than one so that the cube in equation (6.22) is also considerably less than one. As Figure 6.6 shows, however, the range of validity extends out to about one-quarter of a radian or about 15°.

Linearisation using Taylor's expansion

A function $f(x)$ can be evaluated at a small interval h in x by evaluating the differentials. This is a mathematical fact; there is no physical significance to it other than in its application to a physical problem. Thus:

$$f(x_0 + h) = f(x_0) + h\frac{df}{dx} + \frac{1}{2}h^2\frac{d^2f}{dx^2} + \cdots \tag{6.27}$$

For many functions it is not necessary to go beyond the second differential. In fact, not only will higher powers of h diminish in importance for small values of h, the higher-order differentials might themselves become negligibly small. If h is chosen small enough so that only the first differential need be used, the function is automatically linearised in h.

We might expect that Taylor's expansion of the angle through which the pendulum swings should give the same result as our previous approximation. Expanding $\sin\varphi$ over a small angle $\Delta\varphi$ gives:

$$\sin(\varphi_0 + \Delta\varphi) = \sin(\varphi_0) + \Delta\varphi\frac{d}{d\varphi}\sin(\varphi_0) \tag{6.28}$$

At the vertical $\varphi_0 = 0$, so $\sin\varphi_0 = 0$. The differential of the sine is the cosine, and at $\varphi_0 = 0$, $\cos\varphi_0 = 1$, leaving:

$$\sin(\varphi_0 + \Delta\varphi) = \Delta\varphi \tag{6.29}$$

As before, the sine of a small angle is seen to be equal to the angle itself.

Taylor's expansion of the potential energy

The expansion of the sine function is equivalent to the expansion of the force, but in other systems an expansion of the potential energy may be more useful. Let's again take the pendulum as an example. In the rest position the potential energy U is a minimum. That is, if the pendulum is moved off the vertical by a small displacement Δl it is constrained to rise and do work against gravity. The gravitational potential energy increases.

At the minimum $dU/dl = 0$, so the expansion must take in at least the second differential if it is to be of any use. This then leaves the potential energy as,

$$U = U(l = 0) + \frac{1}{2}\Delta l^2 \frac{d^2U}{dl^2} \tag{6.30}$$

This looks neither familiar nor linear, containing as it does the square of Δl, but that is because SHM has so far not been expressed in terms of energy. Consider the linear, central force that is a precondition of SHM (equation (6.20)). The work done, dw, in moving a mass a distance ds against this force is simply:

$$dw = F.ds \tag{6.31}$$

and the total work done in moving a distance s is just the integral over the distance, that is:

$$w = \int F.ds = \int (-k \cdot s)ds = \frac{1}{2}ks^2 \tag{6.32}$$

where the negative sign has been dropped for convenience (negative energy has no meaning other than the fact that the integration is performed in a particular direction). Equation (6.32) is identical to the Taylor's expansion to the second differential (equation (6.30)) with $k \equiv d^2U/dl^2$. In other words, the motion is SHM.

The validity of Taylor's expansion

Taylor's expansion is exact, but strictly it contains an infinite number of terms. Where do we decide to stop the expansion in any given situation? The virtue of the Taylor expansion is that it is general and not specific. What the above example tells us is that *any* bound system, that is where the mass is situated at a minimum in the potential energy, will undergo SHM if it is disturbed from equilibrium slightly, provided the disturbance is small enough so that terms in d^3U/dl^3 are not needed. An electron bound to an atom, an atom bound to a molecule, a satellite bound to a planet, and so on are all examples of bound systems in which a perturbation leads to SHM. This explains the ubiquity of the phenomenon, and one can only marvel that such a simple and elegant piece of mathematics has such a widespread correspondence with the physical world.

The expansion to the second differential is equivalent to assuming that the force is linear in Δl, and as with the pendulum there will be a range of values for the displacement over which this approximation holds true. This then is the key. In this case it is necessary to stop the expansion at the third term. In other cases it may be necessary to stop after the second, or the fourth, depending on what is required.

▶ The complex operator

Elegant though the mathematics of SHM is, it can be made even more so through the use of the complex operator. In choosing to use either a sine or a cosine function to describe SHM we are faced with the inevitable consequence that the other function has to be rejected. This seems a little odd. It is a seemingly arbitrary choice that is not justified by the mathematics. Physically we might suppose that the cosine is actually more realistic as it is necessary first to perturb a system before SHM occurs. Thus we would move the pendulum and let it go. The motion therefore starts at the maximum amplitude rather than at zero, and in this case the cosine is more useful. Once the motion has been established, however, it doesn't matter which we use because the choice of $t = 0$ depends entirely on when we decide to start the clock and the relative phase of the cosine and sine functions is unimportant.

The complex operator is a very useful mathematical device for retaining information about the phase within the mathematics and therefore represents a more complete solution. The complex operator j represents motion through 90°. Suppose a vector of length B lying along the positive axis as shown in Figure 6.7. Successive operations by j rotates the vector through 90° to give jB, $-B$, $-jB$, and finally B again. It can be deduced from this that,

$$j^2 = -1 \tag{6.33}$$

and

$$j = \sqrt{-1} \tag{6.34}$$

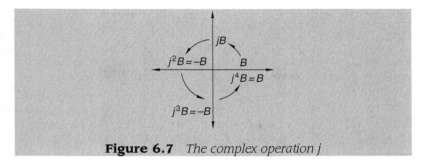

Figure 6.7   *The complex operation j*

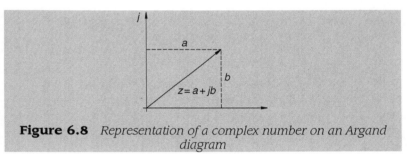

Figure 6.8   *Representation of a complex number on an Argand diagram*

Of course, $\sqrt{-1}$ is not a real number, so a real number operated on by j is called an *imaginary* number. The combination of a real number with an imaginary number is called a *complex* number, and though the notation suggests that the real and imaginary parts can be added algebraically the addition proceeds vectorially. A complex number has two components at right angles to each other, and their graphical representation is known as an Argand diagram (Figure 6.8). The magnitude of this number is therefore given by Pythagoras' theorem, $z = \sqrt{(a^2 + b^2)}$. In other respects the algebra of complex numbers is identical to the algebra of real numbers, as described in Appendix 1.

The complex exponential in SHM

The rotating vector A in SHM has components $A\cos\theta$ and $A\sin\theta$ respectively, so it is possible to write:

$$A(\theta) = A\cos\theta + jA\sin\theta \tag{6.35}$$

It is straightforward to show, using the series expansion of the exponential, that,

$$e^{j\theta} = \cos\theta + j\sin\theta \tag{6.36}$$

so

$$A(\theta) = Ae^{j\theta} \tag{6.37}$$

The rotating vector can be represented mathematically by the complex exponential. Just as operation by j represents a rotation through 90°, operation by $e^{j\theta}$ represents a rotation through $\theta°$. Mathematically this is a much more complete description than either the sine or the cosine function, as it contains within it the information about both components, each of which is easily identified by the existence or otherwise of the prefix j. As imaginary numbers do not exist in the physical world, we concentrate only on the real part, which is the cosine function. There is now a mathematical reason for choosing one over the other. Making the assumption, as before (equation (6.15)), that θ increases with time, that is:

$$\theta \propto t = \omega t \tag{6.38}$$

leads to the complex exponential description of SHM. This is not only a more complete mathematical description, it allows physical situations to be more easily modelled. Real oscillations do not last forever, as shown in the last chapter. If the oscillation is driven in some manner then of course it will last as long as the driving force lasts, but dynamically this is a different problem. The driving force is an additional force that must be considered in Newton's second law, and the solution is different from that arrived at in the absence of a driving force. Experience tells us that real oscillations die away, and the mathematics of SHM should allow for this.

The sine and cosine functions do not easily lend themselves to this problem, but the complex exponential does. If the frequency ω is written as a complex number, the time evolution of the oscillation demonstrates a decay. For example, let

$$\omega = \omega_r + j\omega_i \tag{6.39}$$

where the subscripts refer to real and imaginary parts respectively. Then,

$$e^{j\omega t} = e^{(j\omega_r t - \omega_i t)} = e^{-\omega_i t}e^{j\omega_r t} \tag{6.40}$$

The complex exponential again represents the oscillation, but now it is prefixed by a term that decreases in magnitude as t increases. Stated like this, the introduction of a complex frequency sounds like an *ad hoc* construction, something that is just added on to serve a purpose. It isn't. The result can be derived rigorously from the dynamics but it is a lengthy process and serves no purpose in the present context.

You may wonder as to the physical interpretation of a complex frequency. There is none. It is a mathematical tool that allows for decaying oscillations to be represented. If you were to measure the amplitude and frequency of a damped oscillation, as the system is termed because there must exist a dissipative force such as friction that damps the oscillation, then you would measure only the real part of the frequency and the amplitude as a function of time. You could of course represent the oscillation as the product of a sinusoidal oscillation and an exponential decay, but this really is *ad hoc*. You would not be able to derive such a function from the dynamics of the motion as represented by the second order differential equation.

▶ Oscillations to waves

The extension of oscillations to waves provides the physicist with an even more powerful mathematical tool. Just as the mathematics of SHM finds widespread correspondence with the real world, the mathematics of waves extends our ability to describe a whole new range of phenomena. In particular, Schrödinger's formulation of quantum mechanics uses the notion of waves to describe matter, so there is a mathematical thread that links a simple pendulum to an electron in an atom or in a collection of atoms, such as a solid, to a laser and beyond.

The first question to ask about a wave is, 'What is being transported?' At first sight it can appear to be matter, but that is because our senses deceive us. Hold a piece of string at one end and move it such that a wave or pulse moves down the string, and it appears very much that the string is moving with the pulse. Obviously this cannot be the case because the string is held at one end. There are more sophisticated ways of doing this experiment but the result is the same. If a mark on the string were to be observed, this mark would execute SHM in the transverse direction. It is really energy that is transported, and to describe it we adapt the ideas of SHM.

Consider two 'particles' on the wave, as illustrated in Figure 6.9. We can describe the displacement by the symbol ψ, so the oscillation has the form:

$$\psi = \psi_0 e^{j\omega t} \tag{6.41}$$

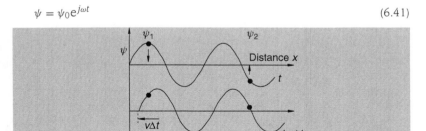

Figure 6.9 *The relative phases of two oscillators at time t and t + Δt*

If the amplitude ψ_0 does not vary with distance, each oscillator ψ_1 and ψ_2 can be described by a similar equation. However, each must have its own characteristic time because the two oscillations are not identical. Oscillator 1 is moving down as oscillator 2 is moving up, as can be seen if the time is moved on slightly so that the wave has propagated to the right. Describing a wave as a collection of oscillators each with its own characteristic time is not very useful. What we want is a way to describe the oscillation at any point x on the wave in relation to some predefined origin.

Let us imagine that all the oscillators are initially at rest and that each only starts to oscillate as the front of the wave reaches it. Suppose we start the clock when the first oscillator ψ_1 starts to move. It will take a time interval,

$$\Delta t = \frac{\Delta x}{v} \tag{6.42}$$

where Δx is the separation between the oscillators and v is the velocity of the wave, before the wavefront reaches oscillator ψ_2. Oscillation ψ_2 is *delayed* with respect to ψ_1 so we can write:

$$\psi_2 = \psi_1\left(t - \frac{\Delta x}{v}\right) = \psi_1 e^{j\omega\left(t - \frac{\Delta x}{v}\right)} \tag{6.43}$$

If ψ_1 is set at $x = 0$, so that $\Delta x = x$, and putting $\omega/v = k$, the wave-vector, then at any point x,

$$\psi = \psi_0 e^{j(\omega t - kx)} \tag{6.44}$$

This is the mathematical form of a wave. Although it has been based on the idea of physical oscillators, it is not necessary mathematically to invoke such oscillators. Indeed, it was the insistence of early physicists that there must be something physical to oscillate that led to the notion of the aether, but electromagnetic waves do not need a medium through which to propagate. This mathematical description is therefore general and finds widespread application throughout the physical world.

▶ The wave equation; partial differentials

Just as the complex exponential description of SHM is a solution to a second-order differential equation, we might expect the wave function ψ above to be a solution to a second-order differential equation as the mathematics is very similar. However, the differential equation now has to represent distance as well as time. If the time-dependent behaviour is observed at a fixed point in space then it will be SHM, as already described. We know this is a second-order differential equation. If the time is frozen the wave equation must also describe the wave motion along the direction of propagation and intuitively we might expect this also to be second order. Therefore, considering only the time dependence,

$$\frac{\partial^2 \psi}{\partial t^2} = -\omega^2 \psi \qquad (6.45)$$

where the notation has been changed to reflect the fact that time is not the only variable, but it is the only variable of interest at the present moment. We call this a partial differential. In partial differentiation, all other variables are treated as constants, and the differentials are curled to distinguish them from exact differentials.

Looking now at the spatial dependence,

$$\frac{\partial^2 \psi}{\partial x^2} = -k^2 \psi \qquad (6.46)$$

These two can be combined by substituting ψ in (6.45) into (6.44) to give:

$$\frac{\partial^2 \psi}{\partial t^2} = c^2 \frac{\partial^2 \psi}{\partial x^2} \qquad (6.47)$$

where $c = \omega/k$ is the velocity of the wave. This is a second-order partial differential equation (PDE).

Electromagnetic waves

James Clerk Maxwell (1831–79) put into mathematical form Faraday's theories of electricity and magnetic lines of force. In his research, conducted between 1864 and 1873, Maxwell showed that a few relatively simple mathematical equations – partial differential equations – could express the behaviour of electric and magnetic fields and their interrelated nature. The four partial differential equations describing electromagnetic phenomena are known today as Maxwell's equations. They first appeared in fully developed form in *A Treatise on Electricity and Magnetism*, published in 1873. These PDEs are not the same as the wave equation, but in the course of his work Maxwell did show that an oscillating electric charge produces an electromagnetic field and that his equations implicitly required the existence of an electromagnetic wave travelling at the speed of light. The wave-like properties of electromagnetic fields were derived by showing that mathematically they obey the wave equation, with a velocity given by:

$$c = \frac{1}{\sqrt{\epsilon \mu}} \qquad (6.48)$$

where $\epsilon = \epsilon_0 \epsilon_r$ is the permittivity and $\mu = \mu_0 \mu_r$ is the magnetic permeability. Heinrich Hertz discovered such waves in 1888.

In the absence of matter, that is in free space, $\epsilon = \epsilon_0 = 8.85 \times 10^{-12}$ Farad/metre and $\mu = \mu_0 = 4\pi \times 10^{-7}$ Henry/metre, so the velocity of light is given by these two fundamental quantities. Maxwell realised that electrons can oscillate at any frequency, so visible radiation – light – is but one small part of a vast spectrum. In the presence of non-magnetic matter, which comprises the vast majority of glasses and crystals to be found in nature, the relative permittivity, $\epsilon_r > 1$ and the relative permeability $\mu_r = 1$, so the velocity of light is given by:

$$v = \frac{c}{\sqrt{\epsilon_r}} \tag{6.49}$$

This might not be immediately familiar, but in fact it corresponds to the phenomenological definition of the refractive index n as the ratio of the velocity of light in vacuum to the velocity of light in a medium, that is:

$$n = \frac{c}{v} \tag{6.50}$$

By comparison then,

$$n = \sqrt{\epsilon_r} \tag{6.51}$$

▶ Optical physics

A complete description of the propagation of electromagnetic radiation and its interaction with matter requires the application of Maxwell's equations. In optical physics, however, specification of the refractive index allows most problems to be tackled. The refractive index determines the amount of light reflected from a surface via Fresnel's well-known relationships, the degree of refraction, the phenomenon of total internal reflection, the action of a lens, and also the penetration depth of the electromagnetic wave inside the material. This last might seem a little odd at first. We associate lenses and other optical materials with glass and crystals, such as diamond, which are transparent. The depth of penetration would therefore seem to be infinite. In fact, transparency is a special case of the general.

The refractive index is defined via the relative permittivity (equation (6.51)), also called the dielectric constant, which is related to the *polarisability* of an atom. This is the extent to which the electronic and nuclear charges in an atom separate when the atom is exposed to an electric field. Essentially we are only concerned with the outer electrons, as these shield the inner electrons from the electric field. Thus, an electric field is applied and the electric charges within an atom separate to some extent in response. If the electric field oscillates, as is the case in an electromagnetic wave, the charge separation (polarisation) must also oscillate. In fact it oscillates with SHM.

An electron orbiting a nucleus constitutes a bound system. The electron is held away from the nucleus by the interaction with the inner electrons but is prevented from leaving the atom by its interaction with the nucleus. It has been shown that any bound system undergoes SHM if it is perturbed slightly, so the outer electron in an atom is just such a system. The electric field constitutes a harmonic driving force with a frequency equal to the frequency of light. The solution to this sort of system has already been described and demonstrates resonance. The polarisation of the atom will exhibit resonance at some natural frequency of vibration, and in consequence the relative permittivity also exhibits resonance. In fact it can exhibit several resonances, and the total permittivity is given by the sum of all these responses. Each resonance occurs at a different frequency corresponding to transitions among different energy levels, and for a comprehensive model of refractive index each of the resonances must be included. However, for the purposes of this argument, where the intention is to understand something of the physics, resonances at higher frequencies can be ignored.

Models of refractive index

Scientific investigation into the optical properties of materials, both crystals and glasses, has not just concentrated on their experimental characterisation. There have, in addition, been many theoretical investigations. In consequence there are numerous models in the scientific literature that describe the optical properties of solids; they are all based on the idea of resonance in forced, damped, simple harmonic oscillators.

As with the *ab initio* model of resistivity described at the beginning of this chapter, it is one thing to lay down the general principles but quite another to apply them to specific solids. Not only must the relative height of each of the resonances be specified, but so also must the resonant frequency and the width. In addition, the presence of nearby atoms in the solid affects the final form of the permittivity because the neighbouring atoms are close enough to affect the potential of an electron. Accepting, then, that a detailed model requires specific modifications, but further accepting that these can be ignored in the first instance, we have, by intuitive rather than rigorous reasoning, for a single resonance in the permittivity:

$$\epsilon_r \propto \frac{1}{(\omega_0^2 - \omega^2) + j\gamma\omega} \tag{6.52}$$

where, from equation (5.6), γ is a constant characterising the strength of the resistance to the electron's motion.

This is the complex-number representation of resonance. It doesn't resemble equation (5.9) for the amplitude of a forced, damped harmonic oscillator, but the maths there had been taken further to eliminate the complex number representation. This can be seen by rationalising the complex number (Appendix 1), that is by multiplying the left-hand side by

$$\frac{(\omega_0^2 - \omega^2) - j\gamma\omega}{(\omega_0^2 - \omega^2) - j\gamma\omega} \tag{6.53}$$

to give

$$\epsilon_r \propto \frac{(\omega_0^2 - \omega^2) - j\gamma\omega}{(\omega_0^2 - \omega^2) + \gamma^2\omega^2} \tag{6.54}$$

If the magnitude of this complex term is taken by squaring the real and imaginary parts, then,

$$\epsilon_r \propto \frac{\sqrt{(\omega_0^2 - \omega^2)^2 + (\gamma\omega)^2}}{(\omega_0^2 - \omega^2) + \gamma^2\omega^2} \tag{6.55}$$

which is recognisable as equation (5.9). Accepting the relative permittivity to be a complex number, then,

$$\epsilon = \epsilon_1 - j\epsilon_2 \tag{6.56}$$

where ϵ_2 is negative because of the rationalisation. Also ϵ_1 and ϵ_2 can be seen to depend on the frequency of the optical radiation.

The complex refractive index

The refractive index is given by the square root of the permittivity, and a moment's thought will tell you that it too must be complex. For example, if the refractive index is written as:

$$n = n_r - jn_2 = n_r - jk \tag{6.57}$$

where the notation k, commonly called the extinction coefficient, is introduced for the imaginary part of the refractive index n_2, it is possible to show by the inverse process of squaring equation (6.57) that,

$$\epsilon_1 = n_r^2 - k^2 \quad \epsilon_2 = 2 \cdot n_r \cdot k \tag{6.58}$$

The mathematics is not rigorous, but it doesn't need to be. Intuitive processes such as squaring (6.57) are perfectly acceptable. The complex nature of the refractive index is thereby confirmed and that is the important point.

By a process of intuitive argument, then, backed up by a little bit of mathematics we have established that the refractive index is complex, but what does this mean in reality? Returning to the complex exponential formalism for a wave, and substituting for the velocity c/n, it is clear that:

$$\psi = \psi_0 e^{j(\omega t - n k_0 x)} \tag{6.59}$$

where the wave-vector is now written as k_0 to distinguish from the extinction coefficient. Substituting the complex refractive index into this equation gives:

$$\psi = \psi_0 e^{-k k_0 x} e^{j(\omega t - n_r k_0 x)} \tag{6.60}$$

which is the familiar exponential decay. As with a complex frequency in SHM a complex refractive index represents loss. The amplitude decreases exponentially as the wave progresses in the x direction.

The physical interpretation of this is that if the frequency of the electromagnetic radiation is close to the resonant frequency, the energy of the electromagnetic wave is absorbed by the medium through which it is propagating as it penetrates further into the material. In the discussion on the forced, damped harmonic oscillator in Chapter 5, it was stated that one of the hidden implications is that at resonance energy is absorbed by the oscillator from the driving force. The outer electrons of the atoms interacting with the electromagnetic wave undergo resonance and thereby absorb and dissipate the energy of the wave, causing the amplitude to decay exponentially with distance travelled into the material.

The refractive index and transparency

Absorption of light, as represented by the extinction coefficient, does not occur over the whole range of frequencies in the electromagnetic spectrum. Consider the most basic form of the permittivity (equation (6.52)):

$$\epsilon_r \propto \frac{1}{(\omega_0^2 - \omega^2) + j\gamma\omega} \tag{6.61}$$

which can be rationalised to make the real and imaginary parts explicit (equation (6.54)). If $\omega \ll \omega_0$, such that $\gamma\omega \ll \omega_0^2$, then $\epsilon_1 \gg \epsilon_2$; that is the imaginary part is so small as to play no significant part. Hence, the imaginary part of the refractive index, the extinction coefficient, is also negligibly small. As the exponential decay of the intensity, which is given by the square of the amplitude, depends directly on the extinction coefficient, there is effectively no absorption of radiation and the refractive index appears to be virtually independent of the wavelength. This is the situation commonly found in glass. The refractive index depends very slightly on wavelength, otherwise there would be no dispersion in a prism for example, but it is a small dependence, and there is no electronic absorption of the visible radiation.

▶ Exponential decay: a universal phenomenon?

So far three examples of an exponential decay have been presented: the decay over time of a damped harmonic oscillator, the decay of the number of radioactive ions, and the decay over distance of an absorbed wave. Do these have anything in common?

On the face of it, no; not even the wave and the harmonic oscillator. Both involve harmonic oscillators, of course, but one *is* a harmonic oscillator whilst the other simply interacts with them. There is no reason to suppose that the time dependent decay of one is directly connected with the distance dependent decay of the other. There is, however, a mathematical connection. Wherever the complex exponential representation of oscillatory phenomena is encountered, dissipation is represented by a complex coefficient. Even if there is no direct physical link there is a common mathematical theme.

What of oscillations that are not simple harmonic? Fourier showed that any time-dependent motion could be expressed as a sum of harmonic components, so each of the harmonic components will decay exponentially. The net effect may not appear to be an exponential decay, but eventually it will approximate to that as components die away at different rates. Another way of looking at the problem is to imagine an oscillation, like the pendulum, that is not simple harmonic in nature and therefore not described by the complex exponential. The decay will therefore not appear exponential, but as we have seen, any bound system undergoes SHM if the perturbation is small enough. It is only a matter of time therefore before such an anharmonic oscillation decays to an amplitude where the oscillation approximates to harmonic. Then, the decay will be exponential.

The connection between radioactive decay and harmonic decay is less obvious, but it is not so much mathematical as physical. The decay of the wave is expressed in classical terms, but if we invoke the quantum picture of light we must perforce interpret the decay of the intensity as the destruction of photons. This is a statistical process. Assuming one photon does not interact with another, and at light levels normally encountered in the laboratory this is the case, the exponential decay follows from the same physical argument as employed in deriving the mathematical model of radioactive decay. If the distance over which half the photons are absorbed is identified, the remaining photons effectively constitute a new sample and half of these will be absorbed over the same distance.

In any physical situation where the process of destruction of 'particles', be they radioactive atoms, photons or whatever, is independent of the number present, the decay will be exponential. If somehow the process is dependent on the number of particles present, there must be some interaction between them. As with the argument above, once a sufficient number have decayed the possibility of interaction between the remaining particles will be greatly reduced. Then the destruction process becomes independent of the number present and the decay becomes exponential.

In short, even if a decay does not appear exponential initially, the final decay to zero will be exponential. The exponential decay appears therefore to be a universal phenomenon, hence it appearance in so many different physical situations. There is a commonality of ideas that unites a diverse range of physical phenomena.

Differential equations again

As demonstrated throughout this chapter, differential equations are to be found everywhere in the mathematical modelling of physical phenomena. Even though the examples given here are limited, reference has been made to others, such as Maxwell's equations. Differential equations, whether in partial form or ordinary form, are one of the key mathematical tools used by the physicist. By corollary there are a range of integral techniques used by physicists. You will encounter at least some of them within your studies. Sometimes they will appear abstract and unconnected with the physical world, but they do have applications, and the physics is contained within these.

One reason why the differential equation is so common in mathematical modelling is the interconnectedness of ideas. Some simple concepts can be found to apply to a great many situations. Consider two such examples, the *divergence* and *gradient* operators.

▶ Differential operators: divergence and gradient

The notion of an operator has already been mentioned in relation to complex numbers. We use the term to represent a mathematical operation that has some physical significance. Both the divergence and the gradient operator use the same symbol, ∇, to represent a differential with respect to a spatial coordinate. In one dimension, say x,

$$\nabla = \frac{d}{dx} \tag{6.62}$$

And in three dimensions:

$$\nabla = \left(\frac{\partial}{\partial x} + \frac{\partial}{\partial y} + \frac{\partial}{\partial z} \right) \tag{6.63}$$

where the use of the partial differential indicates that the variation in all three dimensions is required separately. This is a mathematical shorthand. In full, the gradient operator, sometimes written as $\mathrm{grad} f(x,y,z)$, where $f(x,y,z)$ is any function of these three coordinates, is written as:

$$\nabla f(x,y,z) = \frac{\partial}{\partial x} f(x,y,z) + \frac{\partial}{\partial y} f(x,y,z) + \frac{\partial}{\partial z} f(x,y,z) \tag{6.64}$$

The three terms on the right-hand side of (6.64) are vectors directed along the x, y and z directions. If you are familiar with the unit vector notation $\underline{i}$, $\underline{j}$, and $\underline{k}$, you can prefix the terms with these unit vectors, otherwise just accept that a direction is implied.

The divergence operator, sometimes written as $\mathrm{div} f(x,y,z)$, has exactly the same mathematical form. It is a differential operator that tells us to differentiate the function. Physically, however, the grad and div operators represent very different things. The divergence is a vector and is written as $\nabla \cdot f$, whereas the gradient is a scalar and is written simply as ∇f.

The physical interpretation of grad and div

The only mathematical difference between grad and div is the dot product; both tell us to differentiate the function. The difference lies in the physical situation that either represents. Both are vector operators. That is to say, both are an essential part of the language and mathematics of calculus of vectors. Vectors, let us remind ourselves, have both magnitude and direction. A field is a vector. It is derived from some source and its direction is determined by the position of that source. An electric field arises from the presence of charges and is directed toward them; likewise a gravitational field arises from the presence of mass and is also directed toward it; a magnetic field arises from the presence of current flow but is perpendicular to the direction of current flow. However, fields can be represented as the gradient of a potential. An electric field E, for example, is written in terms of the potential V as:

$$E = -\frac{dV}{dx} = -\nabla V \tag{6.65}$$

Potentials are scalar. They have no specific direction. Move a mass away from the centre of the earth in any direction and the work done, in other words the gravitational potential energy, depends only on the distance moved. In other words, the grad operator is a *scalar that acts on a scalar to produce a vector*.

By contrast, the div operator is a *vector that acts on a vector to produce a scalar*. As an example let's consider the spatial differential of an electric field E. In Figure 6.10, three cases are shown. The strength of the electric

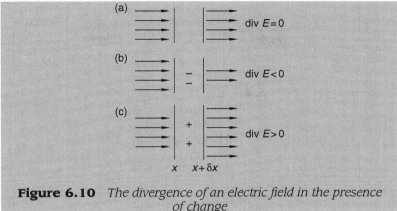

Figure 6.10 *The divergence of an electric field in the presence of change*

field is represented by the number of arrows, and in case (a) four arrows enter at x and four leave at $x + \delta x$. Therefore, the change in E over the distance δx is zero; that is, the divergence is zero. In case (b) the number of arrows leaving is two so the divergence, that is the change in E over δx is negative. Similarly in case (c) there are six arrows leaving so the divergence is positive. We know from simple electrostatics that electric fields arise from positive charges and terminate on negative charges; therefore, if the divergence is negative, as in case (b), there must be some negative charge present to terminate some of the electric field. Similarly, if the divergence is positive there must be some charge present to give rise to the additional electric field. In fact,

$$\nabla \cdot E = \frac{\rho}{\epsilon_0 \epsilon_r} \qquad (6.66)$$

where ϵ_0 and ϵ_r are the permittivity of free space and the relative permittivity respectively and ρ is the charge density. The density of charges is a *scalar*. It has magnitude but there is no direction associated with it.

Continuity equations
The two operators grad and div are very simple but very powerful tools for modelling physical phenomena. The div operator especially lends itself to the problem of continuity. In the example above, where div$E = 0$ the electric field is said to be continuous across δx.

A host of other physical situations can be modelled in the same way. In semiconductor physics the continuity of the electric current is modelled by treating the current flow as a vector in the same way. It has magnitude and direction and as current flow is nothing more than the flow of charge it follows that if the current leaving an element of space δx is not the same as that entering it then some of the charge must stay within the element. Therefore, if the current density is J,

$$\nabla \cdot J = \frac{d\rho}{dt} \tag{6.67}$$

$d\rho/dt$ is the rate of change in charge density.

Heat flow can also be modelled in the same way. If the heat flux entering an element δx is different from that leaving it, then the difference must remain in the element to heat it. If the heat flow is J_q, then

$$\nabla \cdot J_q = -mC\frac{dT}{dt} \tag{6.68}$$

where m is the mass of the element, C is the specific heat capacity, and dT/dt is the rate of change of temperature with time. The rate of change is used in both (6.67) and (6.68) because the flux is the amount flowing per unit time.

▶ The diffusion equation

The continuity equation on its own is only part of the story for a useful model. It tells us what happens in the event of a divergence in the flux but it doesn't tell us what form the flux should take. We need another equation for that. Very often diffusion is the dominant mechanism. Thermal diffusion, like any other form, requires a gradient to drive the process. In this case it is a temperature gradient, and the rate of flow of heat is given by:

$$\frac{dQ}{dt} = -k\frac{dT}{dx} \tag{6.69}$$

where k is the thermal conductivity. In short, heat flows from the hot to the cool, that is in the direction of a negative temperature gradient.

On its own this equation is of limited use. It tells us what the heat flow will be in the event that a given temperature gradient exists, but it doesn't tell us what the temperature gradient will be in any given circumstance. For example, if heat is applied to a surface heat will diffuse into the solid causing a temperature rise. We need to combine both the continuity equation – which gives the rate of change of temperature – with the equation for the rate of heat flow in order to begin to have a sensible idea of temperature profiles. Taking the differential as required by the divergence operator,

$$\frac{d}{dx}\left[\frac{dQ}{dt}\right] = -k\frac{d^2T}{dx^2} = -mC\frac{dT}{dt} \tag{6.70}$$

yields:

$$\frac{d^2 T}{dx^2} = \frac{1}{D} \frac{\partial T}{\partial t} \tag{6.71}$$

The partial differentials are used in recognition that the temperature, T, depends on both position and time. D is known as the diffusivity, in this case the thermal diffusivity.

The diffusion equation – a second-order PDE – is applicable to many diffusion phenomena. In the above example heat is diffusing and causing a temperature rise, but it might be atoms in a gas, defects in a solid, chemicals in a reactor, and so forth, in which case the temperature T is replaced by some sort of concentration.

▶ Boundary conditions

Identification of the conditions prevailing at the boundaries of the problem are often crucial in determining the nature of the solution. As an example, consider diffusion; not temperature this time, but atoms. Suppose we have a solid into which we wish to diffuse some impurities, for example to make a *p–n* junction in electronics. The second-order differential diffusion equation describes the process, but there are a variety of ways that the diffusion can be performed. The difference is all down to the boundary conditions. Consider just two:

1 An infinite supply of atoms at the surface.
2 A fixed supply of atoms at the surface.

In the first case there is no limit to the number of atoms that can diffuse. The situation might be brought about by immersing the solid in a liquid or a gas containing the impurity atoms. The only limit to the surface concentration is the solid solubility of the impurity atoms. The solid solubility describes the maximum number of atoms that can effectively be *dissolved* into the solid whilst still ensuring that the solid remains essentially in its original form. The solid solubility will typically be much less than 1 per cent. If some sort of alloying process, or a chemical reaction were to occur, more atoms might be incorporated but it becomes a different solid in the process, which requires a different model.

The second case may be brought about by depositing a very thin film, just a few nano-metres thick, on to the surface. Once all the atoms in the film have diffused into the surface there are no more at the surface to replace

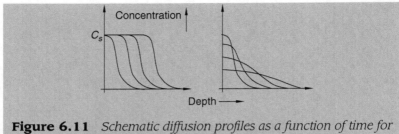

Figure 6.11 *Schematic diffusion profiles as a function of time for different boundary conditions*

them. Atoms may continue to diffuse into the interior to a greater depth due to the concentration gradient, but the surface concentration must decrease in the absence of replacements. Integration under the diffusion profile must yield the number of atoms in the original source.

These two examples are illustrated schematically in Figure 6.11. These particular cases can be solved analytically and the equations of the solution describe the differences between the two profiles. It is worth emphasising again, the essential physics is not different; the boundary conditions are, and in consequence so are the solutions.

► Numerical solutions

Not all differential equations can be solved exactly, and in such cases it is necessary to solve them numerically. As with other techniques mentioned within this chapter, numerical solutions of differential equations is a subject in its own right, so an appreciation of the possibilities is all that can be gained here. Furthermore, the subject of numerical solutions of differential equations is not likely to be taught in the first or second year, but it is useful to know that such techniques exist and can be used.

Let's take a simple example of radioactive decay. We want to be able to calculate the concentration of radioactive atoms at any time t. N_0 and λ (see equation 6.10) must be specified at the outset. This is one of the major differences between numerical and analytical solutions: *analytical solutions are entirely general, provided the assumptions of the model are obeyed, whereas numerical solutions are specific to a set of circumstances*.

The method is to calculate a set of N_1, N_2, N_3, and so on, at times t_1, t_2, t_3 etc. Starting with N_0 at t_0 we want to calculate N_1 at t_1 and so on. Expressing the differential as an approximation, we have:

$$\frac{dN}{dt} \approx \frac{N_2 - N_1}{t_2 - t_1} = \frac{N_2 - N_1}{\delta t} \tag{6.72}$$

Therefore, the model of radioactive decay becomes:

$$\frac{N_2 - N_1}{\delta t} = -\lambda N_1 \tag{6.73}$$

It is now possible to express N_2 in terms of N_1 thus:

$$N_2 = N_1(1 - \lambda N_1 \delta t) \tag{6.74}$$

By repeating the process N_3 can be determined from N_2. This is a form of *numerical integration*. The number of atoms $N(t)$ is being calculated directly, whereas mathematically this can only be calculated after integration (equation (6.12)) and subsequent manipulation.

Second-order differentials can be integrated in a similar manner. Expressing the differential of a function f in the most general form:

$$\frac{df}{dx} = \frac{f_2 - f_1}{\delta x} \tag{6.75}$$

where the variable here is x but could equally be any other variable, such as t, then:

$$\frac{d^2f}{dx^2} = \frac{\left[(f_3 - f_2)/\delta x - (f_2 - f_1)/\delta x\right]}{\delta x} = \frac{f_3 - 2f_2 + f_1}{\delta x^2} \tag{6.76}$$

Thus f_3 can be calculated from knowledge of f_1 and f_2.

▶ Summary

This chapter has been concerned with the process of expressing a physical reality in mathematical form. Differing levels of models have been described, along with different techniques. Some are general and some are very specific, and some, like the numerical methods, are part of the computational revolution taking place in physics. Many of the models chosen will be familiar, but some have been chosen to demonstrate the ubiquity of the mathematical ideas. Familiarity with these makes the process of understanding new models much easier. Whenever you encounter one of these mathematical descriptions you will recognise the physicality being represented and you can focus on the physics and whether it matches the mathematics. The key to both understanding and formulating models is to recognise that a mathematical model represents a physical reality. Both the mathematics and the physics have to be correct, but if the physics is wrong the mathematics is a waste of time.

7 Presenting Scientific Information

The preceding chapters have described the design and execution of an experiment, the treatment of experimental data, and the principles of the formulation of a theory. This is the major part of the research process identified in Chapter 1. There is only one thing left to do; to report on the work. This can be done either in writing or orally, but either way it is an important aspect of the work and in many ways it is as important as the work itself. Science is a public activity, even at undergraduate level, and work must be presented either to your peers or to your lecturers. If you fail to make the most of these presentations you are likely to lose marks. Later, as a professional physicist you could perform the most brilliant experiment but have it ignored by your peers and colleagues simply because you have not reported it well. In order to do yourself justice, therefore, you must be able to present the results of your work in an effective manner.

Oral presentations and written reports have much in common, especially in the organisation of the material and in the need to reflect the critical thinking skills you have exercised in the execution of your work, but they are different and will be taken separately in the following pages.

▶ Written reports

Although the focus here is on undergraduate reports, it should be borne in mind that reports are written for a variety of reasons: a scientific project might need to be written up for examination; an industrial sponsor might require a final report on a project; or you might simply wish to publish a scientific paper. Each type of document will be intended for a specific readership and will therefore have specific features that other documents lack, and it is of course impossible to describe the different formats in any sort of detail. Whatever the type of report, the important point is to organise the information and set it out in such a manner that your message is communicated effectively. Effective means in this context that:

- the work can be read easily;
- it can be understood by people with the appropriate scientific background; and
- it says what you wish it to say.

The undergraduate laboratory report serves as a vehicle for familiarising yourself with the common characteristics of report writing.

The purpose of the report

Before setting out in detail the construction, style and layout of a typical report it is constructive to ask what a report is expected to achieve. Throughout this chapter reference will be made to the readers of your report and how you must consider their needs. This is not so easy to understand. It confused me greatly when I first read this advice many years ago, and in retrospect I can see why; I had no idea, and could not conceive, what a reader might expect from a report. It should have been straightforward enough to put myself in the place of a reader, but it didn't seem so obvious at the time. I simply took the view that I had to write up a piece of experimental work I had performed, and set about putting all the details down on paper. It didn't occur to me that the reader, might not understand what I had done.

This view separates the process of writing the report from the experimental or theoretical work itself, but the report should not be considered as a separate activity. Rather, it is the end result of the investigation. Some might view it as an unpleasant but necessary chore, but it is a skill to be acquired and represents a large part of your own personal development. Appreciate this and you can begin to appreciate what the reader expects from your report, and structuring what you write to achieve your desired aim of communicating your ideas and findings becomes a relatively simple matter.

What the reader wants

Your reader will expect to understand not only what you have achieved and how you did it, but also what you set out to achieve, and why. Most importantly, your reader will expect to *learn* from the report. This means that your reader will appraise and evaluate the report critically and on this assessment will decide whether to accept what you write or dismiss it. Your reader will want to come to a firm conclusion about whether the experiments have been well thought out, whether the underlying theory is sound, whether the experimental technique is sound, whether the results are credible, and ultimately, whether the conclusions are correct.

Your objectives

Your objectives in writing the report are defined by this list of expectations from the reader. As a writer you must structure the report so that the research process and the critical thinking skills employed by you are clear. Not only must the scientific method be good but the report must also communicate effectively your intended message. It should be your aim to provide a coherent account of the work you have done:

- what you set out to do;
- why;
- how you did it;
- what you achieved;
- what does it all mean.

Critical thinking

The above set of objectives is just another way of describing the research process set out in Chapter 1. The research process and the associated critical thinking skills relate only partially to undergraduate experiments, however. The problem will already have been defined in the laboratory script, and to a great extent so will the experimental method, but in order to do the experiment well you are still required to:

- gain knowledge (by background reading);
- understand the problem;
- apply that understanding in setting up the apparatus;
- collect data;
- analyse the data;
- synthesise, that is draw conclusions from your work; and
- evaluate.

Finally, of course, you will write a report on the experiment. It is important to realise that, just as you have applied these skills to your experiment, your reader will apply the same critical thinking skills when reading the report.

The message

In writing a report you must take the global view. You must consider the whole of your message and not just one part. A very common mistake made at undergraduate level is to place too much emphasis on the result of the experiment, but at research level there often is no 'answer'. There is only an experiment and the outcome of that experiment. The difficulty at undergraduate level often arises because you are required to measure a well-known quantity, for example the acceleration due to gravity, and there is an 'accepted' value for it. This does not necessarily mean that your experiment will yield this value, or even a value close to it, though of course you hope you won't be too far away. This hope often leads to a desire to achieve the 'right' result to the extent that the experiment and the report can become wildly distorted. If you stop for a moment and consider the nature of research you will realise that there are no such things as accepted results. If there were there would be no need to conduct research in whatever particular field you might be considering. There is only the experiment and the outcome of the experiment.

Ideally you should adopt the same attitude at undergraduate level. You should imagine that those who mark your report will judge it critically and make an assessment of the scientific worth of the report based not on whether you have achieved the 'right' result but on how well you have conducted the experiment, analysed the data, and drawn your conclusions. In order to do that, though, the report must be readable and communicate your intended message. Therefore, concentrate on

- understanding,
- competence,
- communication.

Do not be obsessed with the actual value of a quantity you derive from your experiment.

Grammar and style

The word 'style' can mean both the format of the report and the way you choose to write. The latter is emphasised here with the layout being left to a later section. The style of the written English is very important. A poorly written report will not impress the reader, and it will reduce the impact of what you say. It is important therefore to pay attention to this aspect of report writing.

Many are the common faults that are likely to obscure your message in some way; some are technical, some are concerned with punctuation and the construction of sentences, and some are concerned with the expressions used to convey ideas. Adopt a natural style of writing that reflects the way you speak. Very often authors adopt a stilted language because they feel that this somehow reflects the serious and logical way that science should be conveyed. Scientific writing, like any form of writing, should be natural, though some formality – possibly discipline is a better word – is necessary to avoid an overly familiar or chatty style. A natural style can, by its rhythm and flow, communicate its intended meaning even if there are grammatical errors.

Spending time working on your grammar might not seem very attractive, but it will be beneficial in the future. Even if you are able to convey your message effectively despite making grammatical errors, think how much better it would be if your English were correct. There might be occasions, however, when you want to alter your style. Perhaps your report will be critical of someone you know will read it, and you want instead to adopt an evasive or diplomatic style. This is acceptable. You are adopting a style for a particular purpose rather than being merely pedantic

Practice in writing

If you have a preconception of what is acceptable in scientific writing, then try the following exercise. Write a short article, no more than two or three hundred words in length, in your adopted style and then rewrite it naturally, jotting down the sentences as you speak. Read them both back to yourself and see which feels better. If you are not sure, ask a close friend to listen and seek their opinion. More than likely, you will find that the natural style is the best, but it is always important to revisit what you write with a view to improving it. Winston Churchill is reputed to have written to a recipient of one of his memoranda, 'I am sorry the memorandum is so long I didn't have time to write a shorter one'. Disciplined and effective writing takes time to achieve.

Common grammatical mistakes

Failure to write a proper sentence. There are several reasons for this. Take as an example the heading of this section; 'Failure to write a proper sentence' is itself not a proper sentence. A proper sentence normally has an object, a verb and a subject; somebody or something must do something to somebody or something else. I can say, for example, 'You have failed to write a proper sentence', because it is clear that 'You' is the object of the sentence, 'have failed to write' is the action, and 'proper sentence' is the subject of the action. However, 'Failure to write a proper sentence' is not a proper sentence, but it is a heading here and its deficiency as a sentence doesn't matter.

Separated participles. The misuse of participles is another very common mistake which can lead to the creation of improper sentences. The present participle ends in -*ing*, for example *walking, sitting, being*, and its use is normally straightforward. Sometimes, though, the noun attached to the participle is separated from the participle by a full stop rather than a comma, so that two sentences instead of one are created. For example, 'The laser light was not detected. The filter being green'. It is obvious what is intended by this, but it is not well-written. Instead, you might have, 'The filter, being green, absorbed the laser radiation'. The participle 'being' is now much more clearly associated with the filter which absorbed the laser radiation (a more precise term than 'light' and therefore to be preferred) because it is – quite properly – in the same sentence. 'The filter being green' is not a proper sentence. On its own it says nothing; it is a sub-clause, and as such must be separated from the main clause by commas. These mistakes notwithstanding, the construction is clumsy and not one that I would willingly use. Consider instead, 'A green filter absorbed the radiation and nothing was detected.'

Dangling participles. Very often the rule that a participle must have a noun associated with it is broken in technical writing. You will find sentences such as, 'Having measured the current, the resistor was changed'. What does the 'having' refer to? The author, of course. Correctly written, the sentence would read, 'Having measured the current, I changed the resistor'. You might not like the use of the personal pronoun 'I', having been told throughout your school life and beyond that the impersonal, passive sense is preferred in technical writing. You might be right. Personally, I don't find that particular sentence offensive. It is accurate, correct English, and describes perfectly well what happened. If you prefer to stick with the impersonal, though, try to avoid the use of 'dangling' participles and stick to something fairly direct; 'The current was measured for each value of resistance used'. Short, and to the point, and side-steps the problem of the dangling participle.

Punctuation
Misuse of commas. Commas are probably the most confusing punctuation mark in a sentence. There are many circumstances in which a comma is used, and many when it is not. To describe them all would be arduous, so the basic principles will be described.

A comma provides a natural pause in a sentence. Read the sentence aloud to see where the natural breaks occur, and if there isn't a comma there put one in. Conversely, you might find a comma, where you feel that there is no natural break, in which case you should remove it. Re-read the above sentence very carefully before you move on, and see where the redundant comma can be found. The technique isn't foolproof, but you will be right many more times than you are wrong.

Natural breaks arise for a variety of reasons; the sentence might be very long, in which case you will need to develop your lungs before reading it without any commas; or the sentence might simply contain a sub-clause. Sub-clauses are the additional parts of the sentence that add to the meaning. For example, in the sentence, 'The filter, being green, absorbed the laser radiation', the main clause is 'The filter absorbed the laser radiation' whereas 'being green' is the sub-clause. The main clause is the meat of the sentence; it describes essentially what the sentence is about, *viz*. that light is absorbed by the filter. The sub-clause adds to the sentence by saying that the filter is green. All sub-clauses are separated from the main clause by commas.

Colons and semi-colons. The use of the colon and semi-colon is very confusing for many people. They also provide breaks but they are somewhat stronger than the comma. Clauses delineated by commas tend to flow together quite nicely so it is obvious that, together, the clauses make one complete sentence. Clauses delineated by the colon or semi-colon can appear to be main clauses in their own right, so that the presence of the colon or semi-colon denotes a definite break from one clause to the next. 'Why use one of these instead of a full stop?', you might ask. 'Variety' and 'creativity' are the reply. Very often the full stop is too definite; if two main clauses appear to be very closely linked to each other, then try separating them by one of these punctuation marks. In order, the colon takes precedence over the semicolon, which in turn takes precedence over the comma. If two or more of these are to be used in a sentence then the order of use should be colon: semi-colon; and comma.

Much more could be said, but this is not meant to be a comprehensive text on English. Rather, the aim is to provide enough information to help you get started on the road to writing in a good style. The key to developing a good style, however, is practice. You must take responsibility for what you write, and you must make the effort to develop your own style. Each time you write something, even if it only a scribbled note in your lab' book, ask yourself whether the meaning is clear, whether the punctuation is correct, and so on. You will probably find initially that this leads to your creating some pretty cumbersome phrases, but within a short space of time you will progress beyond that and start to use a good style naturally.

The report itself
There is no precisely defined format for a written report. For anyone to say you should write this or that would be quite wrong, and this probably contributes to the confusion not only of undergraduates but many relatively inexperienced authors. What you write in a report depends on the report itself, so it is vital to appreciate precisely why you are writing a given

document. For example, if you were writing a report on a first year under-graduate experiment, you might wish to include a whole section devoted to the evaluation and the significance of errors in the experiment because you have been instructed to pay particular attention to that. If the report is concerned with a lengthy piece of experimental work, say a final-year project or a research thesis, it would not be necessary to include this detail but you might wish to include details of much of your experimental work as it has developed in order to give the examiner a true impression of the effort you have put in. However, none of these things would necessarily be appropriate to a scientific paper, where you would more properly put emphasis on the results of the work.

The story

What you write depends very much on whom you intend should read the report. It is vital to understand this, for once you grasp this principle you should have little difficulty in tailoring your report to suit the situation. Irrespective of the type of report, whether you write a good report or not depends on how you organise your information. A report must tell a story, with a beginning, a middle, and an ending. Decide first upon your story, and then set about telling it in an appropriate way. This means that you should be very clear about your results and their significance before you start writing, so that you have an idea what it is you intend to write about. Then you can tailor the introduction and the method to lead naturally into the results.

Do not suppose, however, that you have unrestricted freedom when choosing the format of your report. There are time-honoured practices that should be followed wherever possible. A scientific report will normally be subdivided into sections, comprising:

- Title
- Abstract
- Introduction
- Experimental or Theoretical Method, depending on the type of work to be presented
- Results
- Discussion, about the results
- Conclusion.

The title

Choosing the right title for a report is very important. Sometimes the title is the first thing available to a prospective reader who will then decide whether to expend further effort trying to obtain a copy of the work. A title should be precise and accurate and give the reader as good an indication as possible what the report will contain. All this must be balanced against excessive length, however.

One of the most serious errors committed by inexperienced authors is to use a title which is too general. For example, suppose you had done some work on the properties of thin films of silicon dioxide on silicon, you might be tempted to write simply, 'Thin Films of SiO_2 on Si' as your title. Now imagine that you are a researcher with an interest in this field; would you give the paper much attention? After all, you're busy and you want to know before you even read the document that there is a strong chance it will contain something interesting and useful. You would want to know, for example, whether the films were deposited, and if so by what method, or grown by heating the silicon in a furnace in the presence of oxygen (known as thermal oxidation). You would also want to know what property of the films is under investigation; electrical, optical, mechanical, tribological, interfacial properties, and so on. 'Optical Properties of Thermally Grown Thin Films of SiO_2 on Si' is much more informative and not much longer. You should think of the reader at all stages in the writing of your document.

The abstract

The abstract, as the name implies, is abstracted from the report and stands alone. Abstracts of scientific papers are often published on their own, along with the title and the list of authors. Prospective readers of your report might well find only the title and the abstract in a database. The title, as discussed, will catch their interest first, and details not contained in it should be contained in the abstract. In this way, a reader will know whether your report contains anything of interest. Therefore, the abstract should contain a series of statements to say what it is you have done and the conclusions you have come to; *there is no room for discussion or for long descriptions*. A paper will typically consist of a few thousand words but the abstract will normally be less than two or three hundred words. You have to be concise.

Suppose, for example, that you had performed a series of experiments to verify Ohm's law. You have chosen to apply a voltage to a series of tungsten wires of identical length, but of differing cross-sectional areas, and to measure the current flowing in the circuit. On performing these experiments you find indeed that the current is proportional to the applied voltage up to a point, after which the current increases sub-linearly with voltage. You conclude, after modelling the problem, that the current flowing induces sufficient heating in the wire to cause a change in resistance, but there is nothing fundamentally wrong with Ohm's law. How do you write an abstract for this experiment?

Your abstract would not be very different from what's written above. With only slight modification you could adapt the preceding paragraph as an abstract for this experiment. It is concise, precise, and contains nothing superfluous. Your abstract should follow suit. Statements such as, 'Ohm's law states that the current flowing in a circuit...' should be left to the body of the report, if indeed they are necessary at all. Nor should you say, 'The aim of this experiment is...'. There is no room for this sort of introduction to the problem, nor room for a description of experimental apparatus. The abstract should be a simple statement of what you have done and what your conclusions are so that anyone reading it, together with the title, knows exactly what the report is about. Again, think of the reader.

I would add one final point about the abstract. If you have to refer to other work in the abstract, for example you might want to say something like, 'We have used the method developed by Smith *et al.* to test...', then you should include a complete reference in square brackets. The statement then becomes, 'We have used the method developed by Smith *et al* [authors, journal, volume, page number, year] to test...'. Remember, the abstract stands alone so it cannot refer to some other part of the report, including the citations at the end.

The introduction

The introduction is probably the hardest section of the report to write. It is the one section where you can exercise all your creativity because there are very often many different ways you could write an introduction to a topic. This freedom creates problems, though. If you are not careful you will lose sight of your objective, which is to write a concise, informative, and readable piece. Putting in too much information can discourage the reader just as much as putting in too little. The majority of people are very busy in their working lives and have little time for a lengthy discourse, however well-written and creative it might be. The introduction is the start of the report. Not only must it introduce the subject, it must also introduce the report itself. The two are not the same thing, so let me take each one separately.

Introducing the subject

I have referred to a report being like a good story. The introduction to a story must set the scene for what is to come. It goes without saying that you must have a clear idea what the story is before you can write it. Simply launching into an introduction hoping that once you get going all will become clear is a recipe for disaster. Have firmly fixed in your mind the results you will present and the conclusions you will lead up to, and then you can write the introduction to fit in with this.

There is a danger in this approach that you should beware of. You will be starting from a different position from your reader. You may be tempted to put information into the introduction that you *know* will be relevant to a subsequent part of the report without making clear the reason for its inclusion at this stage. You will do this because you want to refer back to it later. You know the whole story but your reader doesn't. Your reader knows only what you have written and will at best see such material as irrelevant and at worst confusing or misleading. If you are going to put such information into the introduction you should also include an explanation as to why it is there. The danger to beware of, then, is that you anticipate not only your results but also the discussion.

Ask yourself, 'What is relevant?' If part of your story is to discuss in detail the meaning of your results and any conflicts that have arisen with published data, this is best left to the discussion. It may be sufficient in the introduction merely to mention past work and its implications without going into detail.

Anticipation

The preceding has highlighted one of the commonest mistakes made by undergraduates; that of anticipation. Anticipation of later sections of the report must be avoided at all costs. It is not a problem confined to the introduction, it can occur anywhere, but the introduction is the first place it is likely to start. In the above, you put something in to the introduction because you specifically want to refer back to it later. In this sense it is a deliberate act rather than a case of ill-disciplined writing which causes you to mention things out of place, but for whatever reason it occurs anticipation has two undesirable consequences. First, the reader becomes irritated because when something is first mentioned out of place all the information necessary to put what you write in context is not available, and second, the impact of what you write is reduced. When the reader eventually gets to the proper place for something that has already appeared earlier there will be a sense of repetition, a feeling that this has already been covered and therefore can be skipped over. The loser in this case is you.

Creative writing

You should be both disciplined and creative if you are to write good reports. Discipline is essential to ensure that everything is in its proper place in the story, and creativity is necessary to ensure that your report also has literary value. Talk of creativity might sound strange to some to whom science appears to be the apotheosis of logical thought. The old science-fiction films of the 1930s portrayed this idea with scenes in which the principal character, often a dedicated scientist played by someone like Boris Karloff, would say something like, 'I'm sorry my Dear, but I must keep working. There is no room for emotion in the life of a scientist'. I haven't made this up; I remember just such a quote from one film in which he then went to his laboratory, there to stay until the small hours of the morning to perform some devilish experiment that transformed him into a human monster. Things are obviously not quite like that in reality, but judging from the style of writing that some people adopt, it is easy to believe that they have taken to heart this nonsense about emotion and scientists. Art and science are, in my opinion, but different aspects of the same. Both are in essence creative – where does the inspiration for a new theory or a revolutionary experiment come from? – and both require the exercise of logic and discipline to bring the initial creative idea to fruition.

I have written about using a natural style rather than some false adoption. Be confident; be creative; be natural. You have performed your experiment, you have analysed the results and determined the physical significance of them, and now you must convince your readers of your arguments. You must construct an argument that is logical, that starts at the beginning and ends at the end, and where, preferably, these two are not too far apart. You must decide where each part of the story fits and construct your introduction so that everything follows logically from it.

Putting the subject in context

You will need to set your subject in context. By this I mean simply that your reader will want to have some idea of what it is *you* hope to achieve from this experiment. Is it a technological advance, additional knowledge, an alternative technique, an attempt to provide definitive data to resolve a dispute in the open literature? There are many reasons for conducting an experiment, and you should be clear about these so you can convey the right message to your readers. At undergraduate level, of course, one of your principal motivations will be simply that you are required to do an experiment as part of a laboratory course. Looked at another way, however, your aim is to understand and to increase your knowledge in a particular sphere through experimentation. Take this as your starting point, then, and treat the experiment as a piece of research. Setting the work in context could then involve a description of the importance of the knowledge (for example, electric motors would be impossible without an understanding of

Fleming's left-hand rule and a report on the latter could be introduced with reference to the former) and a description of other methods which have been used to investigate the effect (with references of course). Whatever you write, the final decision on the content is yours, but you must be clear that it is relevant. Your reader will certainly notice if it is not and again you lose.

It will not always be possible to give reasons for doing the work, especially if there is some commercial sensitivity. Nonetheless, if you can give a good scientific reason for the work then do so. When referring to other work in the open literature which is pertinent to your own, you could include a brief description of the work in question and refer the reader to the original publications, but as described, you should make every effort to avoid anticipating later sections of the paper, especially the results and discussion. The introduction should contain the *minimum* amount of information required to set your own work in context, rather than the maximum. Remember, your reader will be starting from the position of knowing next to nothing about your work so you will want to include sufficient information both to set the aims of the experiment in context and to set the results in context. These two can be quite different.

There are no specific 'dos' and 'don'ts', merely good practice and bad practice. If you feel that it will improve your report to include a lengthy review of the subject in the introduction, then do so. Do not feel compelled to do it, though. Make a judgement on what is best for your report, taking into account what you feel is necessary to present your work in the best light, and the likely reaction of your intended readership. Be disciplined. It is easy to write page upon page in the mistaken belief that you are being informative, only for your reader to think, 'This is all very interesting, but where is it leading'. If you are going to provide a review, then write somewhere your reasons for so doing, so that the reader knows why they are being presented with so much information.

This, then, is how you introduce the subject. Give your reasons for doing the work, if necessary, and hence what you aim to achieve, and provide your readers with enough information to set your work in context. It is probably better to make this your normal practice and only present a lengthy review of the subject when it is strictly necessary.

Introducing the report

I once attended a talk by a fellow from the Ministry of Defence in which the speaker said that it was an MoD custom to tell people things at least three times: tell them what you are going to tell them, tell them, and then tell them what you have told them. You need the first of these to introduce your report. That is, you tell people what you are going to tell them. Consider, for example, the experiment on Ohm's law referred to previously. An introduction to the report could consist of the statement, 'This report presents a series of experiments designed to test the validity of Ohm's law, which show that apparent deviations from the law are not in fact real, but are perfectly consistent with the law.'

You might now be wondering what is the difference between the above and anticipation. This statement does not describe the experiments, nor does it give any results. It merely states what the experiment is about and the conclusion you have reached so the reader knows what information will be gained by reading the report. Nothing really is anticipated; you still have to read the report to understand any of the detail. All you are doing, in fact, is *signposting* the way so that the reader knows what the report is about and does not get lost. A report written without signposts is rather like a mystery tour; it might lead to an exciting climax, but it can be very frustrating on the way. Remember, you are trying to be informative.

It is important to exercise discipline in this part of your report. If you are not careful, signposting the way can easily lead to anticipation. Of course, the less you write, the smaller the probability that you will get it wrong, and whilst it is always a good idea as a matter of principal to keep your report as brief as possible, it should also be emphasised that you must not make your report unnecessarily brief. University reports can be as short as a few pages for undergraduate experiments, forty or fifty pages long for final-year experimental projects, and up to two or three hundred pages for a postgraduate thesis. *For a large report such as thesis you ought to include at least a few lines describing the content of each of the chapters, rather than just a terse statement on the subject of the thesis.*

The introduction to the report is normally best placed at the end of the introduction, but again there are no hard and fast rules.

Additions to the introduction

Sometimes you will read a report in which the introduction contains more than is described above. For example, you might encounter a fairly detailed description of some underlying theory. It is not altogether clear whether this extra information falls into the category of 'introducing the subject' or 'introducing the report'. To some extent it doesn't matter. The distinction between the two has been made principally to make it clear that the introduction must do both of these things, but if both can be done at the same time then all the better. The subject of theoretical treatments will be left until later. Let's concentrate instead on the specific issue of describing the work to be done.

It has already been mentioned as a possibility that you will include information to be referred to later in the report. This is in relation to work that you have done, however. What about work which you were supposed to do, but could not for one reason or another? In some projects, for example, you might have a well-defined brief about which you want to say something. You might have been asked to design a piece of equipment to do a particular job or conduct an experiment to investigate a specific effect. Does a description of this come into the category of 'introducing the subject' or 'introducing the report'? Possibly it comes into both, but you should ask yourself first how much detail it is strictly necessary to include. It will be acceptable in some cases to introduce your subject with a description of the work that you intended to do, but it won't be appropriate in every case. Almost certainly, you will not find a scientific paper written in this style. Most papers describe work that has been done irrespective of whatever was initially intended, and it is probably best that you adopt a similar stance.

If you did everything you intended to do in your experiment then of course there is no problem; you can simply mention in the introduction that such-and-such will be described. As has been stressed elsewhere, the content of the report is your responsibility, and you must exercise your judgement as to what is suitable. Consider then: you have a project where you are going to measure some physical property of a particular system, and you introduce your report with a statement to the effect that you will be preparing samples by whatever method, and making measurements of such and such to arrive at the information you want. So far there is nothing wrong, provided you then don't go on to describe the details of sample preparation or the measurements. Anticipation is an insidious beast. It creeps in the first moment you drop your guard. A general statement of intention is sufficient in this case. The difficulty occurs when the measurements envisaged at the start did not yield the appropriate information and others had to be devised. In some reports I have read the authors decided first to describe the experiments that failed, why they failed, and what they did to overcome this so that they could then mention the experiments that were actually done and which formed the main part of the experimental work. All this was in the introduction! What else was there left to say in the main body of the report?

It doesn't matter whether your experimental programme changes because certain things don't work. If it is necessary to mention anything at all about it, say only something to the effect that your brief is to prepare samples and determine a particular property, and initial experiments will concentrate on such and such technique. On the other hand it might be better to omit this detail entirely. *Your report does not need to be a blow-by blow account; you must decide what the reader needs to know as opposed to what actually happened.*

Theoretical treatments

There are several ways that the theory underlying the work can be reported, and inexperienced authors are often confused as to the best way. Essentially you have two options; include a separate section for the theory, or to mix it in with some other section of the report. It is possible to put some theory into the introduction, but it is also possible to relegate a derivation to an appendix. How do you decide where best to place the theory?

A theoretical treatment is somewhat like a map. It is a set of instructions on how to get from your starting position, which is the set of your assumptions, to your destination, that is an equation or set of equations that describe the physical system under consideration. You need to include in your report enough information for your readers to be able to negotiate the course. This necessarily means that you must think of your readership, as has been emphasised throughout. To continue the analogy, if you wish to tell someone who lives hundreds of miles from you how to get to your house you will have to give fairly detailed information, but if your visitor knows your locale well then you need only give the address. Similarly, if the theory underlying your work is well-known, you need only present the final equations. If it is not well-known, or at any rate unlikely that your readers will be familiar with it, you *may* need to present a lengthy analysis. However, you needn't include irrelevant material; someone travelling from London to Leeds needs to know only that they should travel on the M1, not a description of every mile, every turn-off or every service station. In your mathematics you should present only the essential steps in the argument, rather than every integration or algebraic manipulation. These are acceptable in a text book but not in a report.

Once you have decided how much theory to include in your report, the question posed at the beginning more or less answers itself. If you are presenting only a few equations, you can easily incorporate them into the introduction, as part of the background to the subject. A more detailed treatment will probably warrant a section on its own, lest the introduction become overly long and you lose the reader in detail. If you think it necessary to include a derivation of a particular equation, but think that the derivation itself will distract the reader from the main point of your report, you should give the main equations in the body of the report and refer the reader to an appendix in which the detailed mathematics can be set out.

No matter how you incorporate the theory in the report, there are certain practices you should follow. Equations should be numbered sequentially, the number appearing on the right-hand side of the page. All equations should be numbered, not just those that you will refer to again. When quoting equations or borrowing mathematical treatments from other sources, you must provide a reference to the source material. Make liberal use of references; your mathematical treatment will be as long as you deem appropriate, but as a reader I would prefer it to be as short as possible with all relevant sections referenced. Finally, if you do relegate a derivation to the appendix, make sure that you refer to it in the text.

The experimental method

The most important point you should remember about writing up the experimental method you have employed is that you are not transcribing a chronological record of events. Your log book will contain a historical account of your work, and you have to convert that to a readable, but no less informative, section of a report. You have to act as a filter between the log book and the report and remove all the unnecessary information.

How much you remove depends of course on the type of report. Undergraduate experiments are perhaps the hardest to write up in this regard because often you have to write up an experiment that didn't work as satisfactorily as you might have hoped. The problem doesn't really exist when writing a paper; the manuscript will go to a referee who will comment on the scientific worth of the paper, so if you are writing about an experiment that didn't work the chances are that it will never be accepted for publication. Undergraduate experiments are different, however. You have a limited time to do a particular thing, and for one reason or another you might not succeed. You still have to write the report. How do you proceed?

Incomplete experiments

Let's first establish that there is no universal prescription for dealing with cases such as this. There can be a variety of reasons for the experiment going wrong, and a variety of degrees to which it can go wrong, from merely obtaining poor results to comprehensive equipment failure. You should first discuss the situation with the laboratory supervisor to decide on the best way to proceed: continue as if the experiment had worked; switch emphasis to something that was successful, even if that was not the original aim of the experiment; or abandon the write-up altogether. You should bear in mind that undergraduate experiments are designed not only to teach you some science, but also to provide you with experience of writing up a report. Even an 'unsuccessful' experiment can be a useful opportunity to learn, and you should not necessarily regard it as an unmitigated failure. You might find yourself in later life in a similar position, but where the stakes are higher. Suppose, for example, you have been contracted to perform some work but have not been successful. Do you write, 'Unfortunately due to failure of vital equipment, it was not possible to ...', or, 'If I had more time then I would have done ...'? No, you do not. You cannot afford to, but many people do just this in an undergraduate report.

Of the possibilities outlined above, the last is the easiest to deal with. It might be that you simply have nothing to write about, and no amount of effort will redeem the situation. Hard luck, but of more relevance here are those situations where at least you do have something to write about. Let's take the simplest case, where your results are not satisfactory and don't show what you set out to show, but nothing else appears to be wrong. Proceed as if the experiment had worked, and deal with the results in the appropriate section of the report. Your treatment of the experimental method should be unaffected by results.

A change of plan

If, on the other hand, you were able to do something else within the time available, you have to make the decision either to dwell on the failures or to concentrate on the success. What you do really depends on the reason for the failure. If it is something over which you had no control, for example equipment not working or not being available to you, I would advise that you concentrate on those things that you did achieve. Consult your laboratory supervisor, but be positive. Take the initiative, and make the suggestion of a switch in emphasis yourself, and you'll probably find your supervisor agreeable. There might be other reasons for the failure, however. Some undergraduate projects are linked to research programmes, and you might have been unlucky enough to land a project where the experiments envisaged at first simply did not yield the appropriate information. You might feel in this case that you need to present at least some of this work, either because it represents a substantial part of your experimental effort, or because the fact that the method did not work is in itself useful.

As described, anticipation of the results is the principal danger to avoid here. You can see how it will happen; in order to describe the new experiments you must first describe the old, including the fact that they failed. However, a detailed presentation of the results is not appropriate at this stage. I have said elsewhere that your writing up of the experimental method should not depend on the results; that is the ideal, but the longer and more complicated the report, the harder it is to achieve. Don't adopt a long-winded and difficult style of writing just because you want to avoid mentioning the results. That is counter-productive, as well as unnecessary. Simply be careful not to get involved in a long and detailed description. If you need, and I stress the word 'need' because often it is not necessary, to mention the fact that an experiment was unsuccessful so that you can describe the other experiments you performed, then say so without embellishment. Write something like, 'this experiment failed to yield the anticipated results, as will be explained later [give details of the appropriate section], and therefore alternative experiments were performed'. Then you can go on to describe the other experiments.

Important details

Having decided that you have something to say, you must now set about writing. What sort of things should you include, and how should you set it down? You must make sure you include *all* the relevant information: a description of samples you have used, including size and method of preparation; diagrams of electrical circuits or equipment; make and model of equipment used where relevant; and, finally, a description of the procedure followed. Some of this seems obvious, and some irrelevant, but again you must consider your readership. You don't want a reader dismissing your work as nothing more than a curiosity because you have left out a vital detail. There are more ways than I would care to mention of producing thin films, for example, and each has its own particular characteristics. Suppose you have produced a film which, for some reason, appears to be different from other films produced by the method you have used. Your reader will want to know everything; the vacuum system used, the base pressure in the vacuum, steps taken to minimise contamination, even the size of the sample. Otherwise, there is a chance that your reader will decide that there is not enough information to assess fully the significance of the results, so while they might be novel and interesting, they must remain nothing more than a curiosity. In effect, your results, and your report, mean nothing. Make sure that anyone reading your report has enough information to repeat your work. *You owe it to yourself to maximise the impact of your work.*

Chronological accounts

Inexperienced authors very often provide a blow by blow account of their deeds when describing their methods; 'then I did this, then I did that', and so on. It may be an extreme example, but I once read an undergraduate report where the author described in great detail how he had fixed some components to a board by drilling the board with a 0.8 mm diameter drill only to find that some components were bigger and the holes needed expanding to 1.0 mm diameter. Not only is this detail completely unnecessary, but the author was also telling his readership in effect that he didn't have the foresight to measure the component before drilling the hole. You must strike a balance between being informative and being boring. Therefore, avoid saying the obvious; it is clear to anyone who has any experience of electronics that holes have to be drilled in circuit boards and components soldered, so if it is not your aim in writing your report to provide instruction in these techniques, omit it entirely. If you have to give a detailed description, confine yourself to those things which are not likely to be well-known by your readers, or which cannot be found elsewhere in the literature. Otherwise, make the description as brief as possible while omitting nothing essential.

Let me try to give an example to illustrate the points I have so far made. An example will help to establish not only the right practices, but also those that are wrong, which is at least as important. You will not be able to adopt a good style if you cannot see anything wrong with what you already do. Suppose we return to the experiment on Ohm's law. There are two ways of verifying the relation between current and voltage; apply a voltage and measure the current, or apply a current and measure the voltage. Let us suppose that you use both methods in your experiment, but that you use a constant current source only after you find you cannot control a power supply accurately enough to apply, say, millivolts to your resistor. Using a constant current source, you find you can apply milliamps, or even microamps, and instead *measure* millivolts. A typical write-up from a first or second-year undergraduate would be as follows:

> Before any measurements could be made the circuit had to set be up. This was done by connecting the ammeter in series with the load, and applying a voltage across it, as shown in figure 1. The voltage from the power supply was adjusted and the current recorded. However, it was found that the control on the front of the power supply could not be adjusted accurately to give voltages below 0.1 V, so a second circuit was constructed. This consisted of a current supply in series with the load and a voltmeter connected across the load, as shown in figure 2. A small current was applied to the load and the voltage measured. Then the current was changed and the voltage re-measured. A graph was plotted, which showed a straight line plot of current against voltage.

Ignoring the fact that there are no figures to go with this text, is there anything about this paragraph that can be improved upon? There are things that you might want to include in a proper write-up, such as the make and model of the instruments, but these are secondary here. You might feel that there is nothing wrong with the paragraph; perhaps it resembles something you have written, or have read? The style is certainly impersonal and in the past tense, but is it good practice? No! It falls into the 'blow-by-blow' style of writing referred to earlier, hence the setting up of a second circuit is described with reference to the failure of the first. As I have indicated elsewhere, you are not constrained to write it thus, however accurate a reflection of events it might be. Added to that are a couple of obvious, and therefore unnecessary, statements; '... the circuit had to be set up' and 'A graph was plotted'. When you look at the results you won't need to be told that a graph has been plotted. Finally, there is the anticipation of the results in the last sentence. Here, in the description of the method is the key result, that the current is linear with voltage and hence Ohm's law is obeyed. What is there left to write in subsequent sections of the report. 'But', you might be thinking, 'the final sentence does not necessarily seem out of place'. This is a direct consequence of the sequential style of writing; 'I did this, I did that, I got some results'. Change the style and the problem disappears:

> Two circuits were used, as shown in figures 1 and 2. For voltages of less than 0.1 V dropped across the load, a constant current source was used to supply a current, and the voltage was measured. For larger voltages, a voltage was applied and the current flowing was measured.

It would be difficult to make this paragraph much simpler or shorter. You can see, however, that there is nothing essential in the first version that is not in the second. The style is still impersonal and in the past tense, but the whole 'flavour' of the passage is different. It is in your interests to keep your text to a minimum by the use of imaginative and disciplined writing lest your reader become distracted from the main purpose of the report. It is absolutely essential that you have in mind the story you wish to tell before you set down anything on paper. Only in this way can you be in control of what is written. This probably sounds a little strange, but as the above example illustrates, the style of writing can influence the content as one thing leads naturally into another. You must control the writing, not the other way around, and this can only be done if you spend time thinking about what you want to write and how you want to write it.

Finally, some of the other things that you can do include making use of diagrams rather than long and complicated descriptions, and referring to prior work; for example, 'The optical system has been described else-where...[reference]', so that you don't need to include a description. Remember, *minimise the length and maximise the impact*.

The results

As stressed throughout, clarity of purpose and presentation is very important, and never more so than in the presentation of the results. This is the real heart of the report. The introduction sets the scene and provides the background for the work; the experimental section tells us what you have done in enough detail for others to repeat the work if they so wish; both of these sections lead to here, the results, and everything else, the discussion and conclusion, follows on from here. This is the essence, without which the report means very little. You might be fooled into thinking, however, that the results section is the easiest to present because the information speaks for itself. Don't! There is plenty of scope to make a mess.

In a long piece of experimental work, you might have performed several related but distinct experiments. You will have broken up the experimental description into separate sections so that the reader is not confused; you should do likewise with the results. Do not be afraid to include sub-headings if you feel that these add to the clarity. However, each sub-section should be clearly presented and self-contained.

It is very common for undergraduates to present combined results and discussion. It is not necessarily wrong and you will find many examples of published papers where this practice has been adopted, but I would caution against it. There might be occasions when your results are so complicated that you need to discuss them as you go so that the reader has a reasonable chance of understanding what you are saying, but this is not always the case and probably not even the norm. It is more usual to present the results first and then discuss what they mean, in which case a separate discussion is appropriate.

'Results' only, or a combined 'results and discussion'?

It is of course very difficult to present the results without at least some discussion, and I have no doubt that this will cause you not only confusion but also some difficulty if you take my injunction literally. Don't make the style overly complicated by trying to avoid completely any discussion of the results at all, but beware not to get distracted. You should add just enough to make sense of the results, but leave the wider issues that arise from them until later. For example, you might have performed some experiments to assess the viability of a particular course of action. The wider issue here is which course you should adopt. This might be a straightforward matter if the results speak for themselves, in which case a lengthy discussion is not necessary, but alternatively you might need to argue the point at length. You will draw upon the results but the discussion is clearly separate. Similar arguments apply to experiments performed to verify a theory; present the results first and discuss separately whether the theory is confirmed.

In presenting the results you should minimise the amount of information on offer so that the reader is not bewildered by your work. It goes without saying that you should plot graphs rather than present tables of measured data, but you shouldn't plot every graph that you have ever measured. If you have measured some property, say the current–voltage characteristic of a semiconductor diode, of a large number (for example 50) of samples, you shouldn't present each individual characteristic. Instead, you should select representative samples to present. For the other samples you should choose some salient feature of the characteristics which allows comparisons to be made, such as the current at some particular voltage. Alternatively the salient feature could be something else, for example a gradient, obtained after processing the raw data. Thus you can reduce an overwhelming number of results to a digestible quantity and at the same time prepare the reader for the conclusions that will follow.

In presenting results you should, of course, make mention of errors. It might be sufficient to quote the error associated with a particular value, be it measured or derived (for example the standard deviation on a mean), but it might be necessary in some cases to include a separate section on errors. Such is often the case in undergraduate experiments where an essential part of the laboratory is taken up with analysis of data and the laboratory supervisor will want to see that you have analysed the data correctly. Let's leave a discussion of the treatment of errors to the next section.

The discussion

The discussion is there to focus on the significance of the reported work; whether a theory is confirmed, whether the experiments were successful, possible shortcomings of the experiment, and so on. In short, what do the results mean? In terms of the critical thinking skills this is where you synthesise your conclusions and evaluate your work.

What you write will depend very largely on the type of report. For example, a typical junior undergraduate experiment will usually have a script for the student to follow. If all goes well the experiment will 'work'; that is to say whatever is predicted by theory will be found experimentally. It will be very difficult, therefore, for the average undergraduate to comment meaningfully on the nature of the results and whether a theory is confirmed merely beyond the bald statement that it is. If the experiment has not worked then you will have to try to comment on this, but there are difficulties here. In a well-designed experiment there should not be any serious shortcomings in the methodology so it will be difficult to suggest substantial improvements to the experiment that will result in more accurate measurements and hence better conclusions. Very often you will not know exactly why the experiment has 'failed' and you will be scrabbling around for answers. You are, of course, in a somewhat false position. A professional experimental physicist might normally have to make a judgement on both the nature of the experimental system, which he will probably have had some say in designing, and the implications of the results for the theory under test. Here, however, you know that the theory is well established and the experiment is designed to do nothing more than confirm this in such a way that you might learn something in the process. You cannot question the theory and you know the experiment should work.

There is a very strong temptation in these circumstances to invent something in order simply to have something to say in the discussion. Statements of the sort, 'I might have knocked the table' add nothing to the reader's understanding and invite a reply along the lines of, 'If you don't have enough sense to know whether or not you knocked the table, how can we trust your judgement on anything?' Try to avoid writing something merely for the sake of writing.

The origins of this particular problem lie in the emphasis placed at school, and sometimes at junior undergraduate level, on errors and accuracy. The discussion about errors and accuracy should be based solely on fact and be as quantitative as possible. As indicated in the previous section, a detailed analysis of the errors is not normally needed unless it is either essential to an understanding of the results – for example, if you have used some fancy mathematical technique to extract significant information from noisy data – or if it has been asked for in a laboratory script. Most of the time the discussion will only address the implications of errors which have already been quoted with the results. For example, you might be looking for a change of x per cent in some measured parameter under certain conditions, but you can only measure the parameter to an accuracy of y per cent. You have to decide whether what you have measured is real or not, based on the relative magnitudes of x and y, and if you decide that you cannot measure a change x because the accuracy of your experiment, y, is not sufficient then you have to try to suggest ways of improving the experiment.

In a similar vein, you could consider the example of theoretical predictions being verified or otherwise by experiment. Most theories contain assumptions of one sort or another, and indeed in some cases the assumptions are necessary to allow the problem to be simplified to an extent where a mathematical treatment can be applied. The assumptions, whatever they are, will have consequences for the numerical values of any predictions from the theory. If you have performed an experiment to test the theory you will be concerned with both the accuracy of the prediction and the accuracy of the experimental result; that is, does your experimental result fall within the error of the theoretical prediction and vice versa. Upon consideration of this your conclusions concerning the validity of the theory will be based, so the discussion will address this point first. It might be straightforward so that no real discussion is necessary, or it might be complicated and require considerable discussion. If you decide that the theoretical prediction is wrong you then have to consider the implications of this: do you need a more realistic assumption or is the assumption realistic but the theoretical approach lacking? This is what I mean by the phrase 'significance of the reported work' in the first sentence of this section.

Having said all this, you should recognise that there will be occasions when no discussion is necessary. You should not feel that you have to add a discussion. The report is yours and the content is your responsibility. If you decide that there is nothing really to discuss beyond that which accompanies the presentation of the results then you might decide that this justifies a joint results and discussion section. So be it. Have confidence in your own judgement.

The conclusion

The conclusion to the report will be little more than a summary of what has already been written. It constitutes the third part of the trilogy I mentioned previously, where you tell your readers what you have already told them. At this stage everything of significance will have been discussed, all the results will have been presented and your conclusions already reached. Here you simply pull them together and present them as one, being careful to avoid repetition of what has gone before. In essence the conclusion is little more than a series of statements joined together. Any reader who is pushed for time should be able to pick up your report and, reading your abstract and conclusions only, come to a basic understanding of the relevance of your work to their own.

References

You should not confuse references with a bibliography. A bibliography is included at the end of an essay to direct the reader to other, very general, sources of information, but the works included are not always referred to specifically in the text. A reference, on the other hand, directs the reader to a specific page in a specific work for further details that have not been reproduced in the current text. Thus specific techniques will refer to a source description, and mathematical analyses will refer to derivations. References add legitimacy to your work; not only can you avoid repeating details of work published elsewhere but you can claim to speak with the authority of others.

The most common format for references is to include a number in square brackets at the appropriate point in the text and at the end of the report, in the references section, include the relevant details of that work next to that number so that the reader can locate that work for themselves. It should go without saying that references are numbered as they appear in the text. If a text has been cited early in the work and again much later it should retain the first number given to it, rather than have two numbers. Phrases such as, 'Jones [or Jones *et al.* if Jones is just one of many authors of the cited work] [4] has shown that ...', 'The method of least squares [5] was used to optimise the function ...' or, 'Several attempts to prepare material by this technique have been reported in the literature [2–6]'.

You can see from the above that the reasons for citing other work essentially fall into three categories:

1 Historical perspective – in the introduction to your report you might wish to refer to work that has already been done to set your own work in context.
2 Making use of past work – when dealing with a theory you can quote the relevant result, and perhaps some important assumptions, without giving a full treatment and refer the reader to the source work. If your work is experimental, you might wish to say simply that you have used the such-and-such technique if that technique is standard.
3 Discussion – when assessing the significance of your work you might want to compare your findings with those of other workers or refer to insights developed by others if they are pertinent to the analysis of your work.

At the end of the report there will be a separate section entitled 'References' which will then list the texts you have cited. You must list all authors of a text at this stage and abbreviated forms such as *et al.* are not acceptable. Hence, using a book and a paper as examples, you might have:

[1] *Micro-electronic Devices*, E.S. Yang, McGraw-Hill (Singapore), 1988, p. 99
[2] *Ibid*. p. 150.
[3] 'Rapid analysis of conductance data in metal-oxide-semiconductor measurements using an approximate method'. D. Sands and K.M. Brunson, *Solid State Electronics*, vol. 37(5), 1994, pp. 383–5.

The abbreviation *Ibid* is short for *ibidem*, meaning *in the same place*, and is used for references that follow one another that refer to the same book or paper.

Appendices

Appendices are usually reserved for topics that are essential to an understanding of the report but somehow seem out of place within the report. I have already mentioned placing a mathematical derivation in the appendix and this is probably the most common use of the appendix. The principal points to remember are: use a separate appendix for each topic, and refer to the appendix in the main body of the report. I have come across many undergraduate reports where the appendix merely exists as a self-contained topic that is not referred to or described in any way. What is the point in that? If the material contained in the appendix is needed to understand some point in the report it must be referred to; if it is not then either the report is lacking or the appendix is not necessary.

Figures and figure captions

Traditionally diagrams are included on separate pages at the end of a manuscript. This is not necessarily relevant today because the advent of the personal computer and desktop publishing facilities means that diagrams can be incorporated into written text easily and neatly, much as you would find in a publication. Undoubtedly this makes the report easier to read, as diagrams contained separately at the back of the document require the reader constantly to flip backwards and forwards through the report. However, if it is not easy to incorporate the diagrams into the text then they should be contained at the back in the order in which they appear in the text. It is important to bear in mind that diagrams so incorporated will be reduced and therefore all the lines, symbols, and text must be large enough and bold enough to be read after reduction.

As stated elsewhere, the use of diagrams can help the reader by providing a picture to focus on. The reader can 'see' what is happening and is not confused by ambiguities which might arise in a badly worded description. Furthermore, the chance of ambiguities arising in the text is reduced because the description can be made much simpler. It is important, however, to include only essential diagrams and not to make the diagrams so full of detail that the reader has difficulty interpreting the picture. Remember to label each piece of equipment or component on the diagram.

If the diagram is a graph of results, you must similarly beware confusing the reader. It is important to sift through your results and present the minimum amount of data necessary to put over your meaning. In particular, keep in mind the following points:

- All axes must be labelled with both text and values. The text should include a description of the quantity plotted together with the appropriate SI unit.
- If more than one curve is included in a single figure a key must be provided. Keys can, and perhaps should, be retained from one graph to another where appropriate. For example, suppose you have measured the current–voltage and capacitance–voltage characteristics of a semiconductor diode at different temperatures. You will plot the current against voltage on one graph and the capacitance against voltage on another. Each curve on the graph will be identified with a unique symbol for the data points to show that it represents a measurement at a specific temperature. It is sensible to use the same symbols for the same temperatures in both graphs so the reader can see instantly from the symbols what curves correspond to what temperatures.
- Figure captions should be provided for each graph. The captions should be concise and convey to the reader the content of each graph. The number of the figure should be provided at the beginning of the figure caption, and figures should be numbered sequentially. It doesn't matter whether one figure is a diagram of equipment, another shows a graph, and a third contains a photograph, for example. It is not appropriate to change the numbering system to suit the type of figure, for example *Figure n* followed by *Graph m*. However, in very long reports, for example theses, you might consider it appropriate to label diagrams according to the chapter, for example Figure 1.1, 1.2 and so on, but within each chapter the diagrams must be numbered sequentially.

Tables

On occasions it is appropriate to present a table of results; for example when you have made a number of measurements of one particular parameter (for example acceleration due to gravity) by several different techniques.

However, there is little point in providing a table of data that is included in a graph. The following points should be borne in mind:

- Each table should have a title and a caption.
- Each column of the table should have an appropriate heading, including where necessary the correct units of the quantities to be found therein.
- Traditionally tables have been included at the end of a manuscript, but again there is no reason not to take advantage of modern word-processing technology and incorporate them into the text.

▶ Summary

The main principles of writing a good laboratory report as described above are summarised below:

- **Title**
  - Provides the first indication of content.
  - Should be specific but not too long.
- **Abstract**
  - Stands alone.
  - Contains the essence of the report but no detail.
- **Introduction**
  - Introduces the subject by:
    - setting the work in context and providing any necessary background information,
    - starting from the position of knowing your story,
    - avoiding anticipation of later sections.
  - Introduces the report by providing an indication of the purpose and content.
  - Can contain additional material, such as relevant mathematics and theory, especially background equations.
- **Theoretical treatments**
  - Similar to a road map.
  - Decide upon how much detail.
  - Should it form part of the introduction?
  - Make use of references and appendices.

- **Experimental method**
  - Concentrate on what was done.
  - Be concise rather than chronological.
  - Avoid obvious statements.
  - Avoid anticipating results.
- **Results**
  - Include some discussion as relevant.
  - Minimise the amount of data presented by using and combining graphs.
  - Quote errors.
- **Discussion**
  - What does it all mean?
  - Avoid meaningless discussion of invented errors.
- **Conclusion**
  - Summarise the findings.
- **References**
  - Provide the source details of works cited in the text.
- **Appendices**
  - Refer to appendices in the body of the report.
  - Use appendices for essential information that does not sit easily elsewhere.
- **Figures and figure captions**
  - Label diagrams.
  - Number all figures (graphs included) sequentially.
  - Retain symbols from one graph to another for ease of reference.
- **Tables**
  - Provide a caption/heading.
  - Don't duplicate data presented in graphs.

In addition to the above, bear in mind the following:

- First of all, you should think of your reader. Your reader is not starting from the same position as you are, and you should lead them through the report in a logical and sequential manner.

- This means necessarily that you will have thought of the story to tell. It should have a beginning, a middle and an end. These sections correspond roughly to the Introduction and Method (beginning), the Results (middle), and Discussion and Conclusion (end).
- Take care throughout the report to avoid anticipating later sections. Everything has its proper place.
- You should be creative in your writing. There are no prescribed ways of writing a report and you should bear in mind that you can tailor the format to suit your particular situation. However, you should temper your creativity with discipline.

These are intended only as reminders. Writing a good report requires diligence, discipline and imagination. Above all it requires practice. Use every report you write as an opportunity to practise and you will eventually acquire the essential skills.

▶ Oral presentations

Oral presentations are another form in which scientific information is often presented, and the same essential principles that apply to written reports also apply here. The information must be organised in much the same manner, and you need to follow the 'three-times' rule of:

- Telling your listeners what you are going to tell them.
- Tell them.
- Tell them what you have told them.

You also need to be clear about your story, but unlike a written presentation you need to practise your delivery so that you don't get lost. What can seem like a momentary aberration on your part can appear to be an awkward interlude for your listeners that distracts from what you are saying, even after you have recovered.

Communicating your message
Most presentations you will be dealing with are intended to inform rather than to educate in the way that lectures do. A presentation is a means of passing information on to your peers, so some basics must be understood if you are to make a successful presentation and deliver a message. In particular your audience must not feel:

- bored,
- frustrated, or
- that time would be better spent elsewhere.

It is a important to ensure that not only is the content important, but also that the structure and delivery of the talk contribute significantly to the audience's appreciation. Don't make the mistake of assuming that the information speaks for itself. It rarely does, and it is up to you as speaker to make sure that you deliver your message.

Audience concentration

A listener's concentration will lapse during the talk. If you expect to talk for 45 minutes and have your audience follow every word you will be disappointed. A speaker *must* recognise and take advantage of these lapses rather than oppose them by presenting vital information when no-one is listening. This, then, is the key; use the moments of high concentration to deliver the important details.

How is this done? Fortunately the concentration lapses follow a predictable pattern of:

- high at the beginning;
- rapidly falling off;
- rising again at intermediate points such as the end of one section and the beginning of another;
- falling again; and
- rising finally, albeit briefly, as the listener perceives the speaker to be approaching the end of the talk.

This allows the critically important points to be identified.

Critical moments in a speech

The critical moments of high concentration can be identified as:

- The beginning. Concentration is highest so take advantage by telling your listeners what they can expect from the talk to stimulate their interest. This means holding back any general information until after the signposts have been made clear. This will enable you to take advantage of the high audience attentiveness at the beginning to get the essential content of your message across. For example, you could start a talk with,

    'Today I shall be talking about our experimental investigations into Ohm's law. I shall be talking about four main areas:

    (1) The continuing need for further experimentation into Ohm's law.
    (2) The experimental methodology.
    (3) The results of the experiments.
    (4) My conclusions.

> Starting then with a description of Ohm's law by way of introduction ... etc.'

Or you could start with:

> 'Ohm's law is one of the most widely applied laws in physics today. It is well-understood but there is still a need to conduct experiments on certain aspects of the law [blah blah blah ...]. I shall be describing those aspects in this talk. I will start with:
>
> (1) The continuing need for further experimentation into Ohm's law as an introduction.
> (2) The experimental methodology.
> (3) The results of the experiments.
> (4) My conclusions.
>
> Experiments in Ohm's law are still needed for a number of reasons ... etc.'

The second of these wastes an important opportunity to get the essentials across by concentrating first on largely irrelevant material. It is nothing more than filler; it may seem to make the talk flow and sound coherent, but other than that it does not add anything substantial. The essential stuff is the material contained in the four numbered points. This is when you tell the audience what you are going to tell them – the first of the 'three-times', but this vital information loses its impact when concentration has lapsed.

- The end. Interest is revived briefly. Use this time to remind listeners what it is you have told them to refresh their memory
- Intermediate times. One part of the talk ends and another begins. Interest and attention are revived briefly, so that interim summaries and introductions can be made. Intermediate times can be used to 'bridge' two sections, by saying for example,

> '... these are the main points that a new theory must satisfy. The approach I have taken to ensure that they are satisfied is as follows ...'

Thus one section ends with a brief reminder of the content (the points that must be satisfied) and the new section opens with a statement about what is to come. In this way the listener: knows what to expect from the talk; is reminded at key points about the content of the talk; and is reminded again at the end what the talk has been about. This is none other than the 'three-times' rule.

The importance of the 'three times' rule

The 'three-times' rule ensures that if there has been a lapse of concentration on the part of the listener so that some details are sketchy, then the listener at least knows the main points, having been told them on several occasions, and can concentrate on those. The 'three times' rule thus coincides with the listener's concentration. Rewritten, it becomes:

- feed their expectations,
- elaborate on the message, and
- help them to recollect the message.

In the middle of the talk, when concentration is low, use:

- interim summaries,
- sub-titling, and
- sectional signposts.

Materials

There are two principal materials in oral presentations; verbal and visual. Verbal material is of course the story you intend to tell. It is important to have a coherent and well-structured story, just as in a written report, but the difference is that your words carry less weight than in the written form. In a report your reader can read a passage any number of times to digest the meaning, but in the oral presentation there is only one go at it. If your listener misses what you say then you run the risk of losing your audience entirely. It is important therefore to:

- break up the presentation into discrete sections so that a lapse of concentration in one part does not badly affect comprehension of the whole;
- repeat key points at strategic times; and
- back up key details with visual material so the listener can read or see at the same time.

The visual materials must therefore complement the verbal. It is a common mistake among inexperienced presenters to put up an overhead or slide without referring to it in any detail but to continue to talk about something else. This immediately sets up a conflict in your audience; do they listen or read? They will not be able to do both, so whatever they do, you lose. If you intend to put up visual information it must be part of the story you are presenting and you must refer to it so that what your audience see and hear are but different versions of the same story.

Anything you do that causes a conflict in your audience is to be avoided. You want whatever concentration your listeners have to be directed at understanding your message. If the message has to be interpreted first some of the meaning is bound to be lost, and again you lose. *Let your visual material constitute the main medium through which information is conveyed and let the verbal material guide your audience's attention to the most important visual features.* Diagrams must therefore be simple with the key elements labelled, and you should explain what it is represented in the diagram. Use a pointer to identify key features. Text should similarly be sparse, preferably in the form of bullet points. Mathematics should be kept to the minimum necessary. Very few people can follow mathematical arguments as they are presented, unless they are very simple, so it is essential you guide your audience through the material. Decide what it is necessary for your audience to know and discard everything else as superfluous.

In summary, your visual material must:

- constitute the main message;
- be simple, so that your audience can see instantly the main points; and
- enhance the audience's understanding of the verbal message by reinforcement.

Maximising the effect: the delivery

You have to deliver your address to maximum effect. As with a written report, you owe this to yourself. The following are helpful:

- You must have prepared your material – story, visual aids. In particular do not copy a university lecture. Such a lecture is intended to educate as well as inform and the information is presented in such a manner as to allow you to take notes. Necessarily the information is presented at a slow pace and may be 'built up' to allow understanding. *Do not turn your back on your audience to draw a picture on the blackboard.* All diagrams should be prepared in advance otherwise you will create an awkward and unnecessary break in your talk while you draw and your audience waits for you to finish.
- Use plenty of overheads or slides. The more talking you do the more you are relying on your audience listening attentively to what you are saying. Their concentration will not allow this over the duration of a long talk.
- Don't use too many overheads, however. I would suggest one every two to three minutes. Any faster than this and your audience will feel rushed.
- Don't be afraid to repeat a title already announced. The chances are that you will have already been introduced, but you can do it again.

- You should use a minimum prompt. Don't simply read a long text from a page otherwise the written word will dominate your style. Neither should you rely on the spoken word to get your message across. Back up the spoken word with an overhead or slide, especially when you have facts or figures to relate, so that the audience can read as well as listen.
- Concentrate on animation, liveliness and audience contact. You must develop the knack of projecting your energy.
- You should structure your material. This helps you break down the message and provides key moments for you to restate your message.
- Include only the essential detail. Think very carefully whether your audience needs to understand how you arrive at an equation, or whatever, before you include every detail. Keep detailed mathematics to a minimum.
- Watch the time carefully. Your listeners will become very frustrated very quickly if you overrun.
- Summarise the conclusions on a separate overhead, again so that the audience can read as well as hear.
- Avoid the use of humour unless it is spontaneous and unscripted. There is nothing worse for your own confidence than to have a 'witty' remark fall flat, whilst your audience will not only feel embarrassed they may also feel some antipathy to you. On the other hand, a spontaneously humorous remark is often appreciated and increases the audience's sympathy with you.

▶ Summary

The essential principles of presenting scientific information orally have been given and may be summarised as follows:

- A presentation is intended to inform rather than educate.
- Structure your material to take advantage of critical moments.
- Use the 'three-times' rule.
- Don't rely on the spoken word.
- Be well-prepared.
- Don't set up a conflict between what is seen and what is heard.
- Engage your audience.

Appendices

1 Algebraic manipulation

Algebraic manipulation is the most basic of the mathematical tools required by a physicist. It is increasingly obvious, however, that not all entrants to university-level physics courses have acquired this skill, so this chapter sets out the basic rules. As with all mathematical techniques, practice is required. Mathematics may be regarded as a branch of logic and its arguments are true in a logical sense. You may be able to follow them easily, but that is not the same thing as understanding them and being able to apply them.

The logic of algebraic manipulation can be expressed as follows: suppose we have some equality,

$$y = f(x) \tag{A1.1}$$

Any operation we might perform on y has to be performed on $f(x)$ in order to maintain the equality. Representing the operation, which can be addition, subtraction, multiplication, division, differentiation, taking logarithms, and so on, by O, we have:

$$O[y] = O[f(x)] \tag{A1.2}$$

where the terms that are operated on have been enclosed in square brackets to distinguish the operation from straightforward multiplication by O.

This is a purely logical statement. Either you see the truth of it or you do not, and this is where many students of physics face their first difficulty with maths. As physicists we are used to looking for the physical reality, but maths can be expressed without reference to such. It doesn't refer to any physics for its development, and it need not refer to any physics as it is taught. Of course, the maths can be applied to physical situations, and if these applications can be seen they can be used to help the physicist understand the maths. Use whatever technique you can to help you learn but recognise also that one technique will not apply to all situations. I stress this here because there is a physical analogy to logic of algebraic manipulation that I shall use, but it is important to recognise that there will not always be a convenient analogy at hand and to spend time looking for it will only delay you in your studies. The best way to approach mathematics in my view is to accept the statement, theorem, or principle and to understand it through application to example problems. The problems may be purely mathematical, they may have some reference to physics, but the important thing is to familiarise yourself with the procedures through practice. There really is no other way.

Having said this, the truth of algebraic manipulation can be understood with reference to weighing scales; not the type commonly found in the laboratory or kitchen where there is just a pan for the weight and some sort of read-out to show the weight, but the old fashioned balance where equal weights have to be added to either side to balance the scales. The balance works on weight, not material, so it doesn't matter whether one side contains sand and the other contains gold dust. This is a very good analogy for the equality above; we have two different things that are somehow equal. Clearly if I add something to one pan I must add the equivalent to the other, or the balance (equality) will be lost. If I divide the contents of one pan by half I must do so to the other, and so on. The basic rule, then, is that whatever is done to one side must be done to the other.

Examples

1   The equation of a straight line:

$$y = m \cdot x + c \qquad (A1.3)$$

where m is the gradient and c the intercept, allows y to be calculated at any value of x. How can x be calculated at any value of y?

This is called 'expressing x in terms of y'. Start by isolating the terms in x, that is by subtracting c from the right-hand side (RHS) of this equality. The same quantity must be subtracted from the left-hand side (LHS), so

$$y - c = (mx + c) - c = mx \qquad (A1.4)$$

Then divide the RHS by m, and likewise the LHS:

$$\frac{1}{m}(y - c) = \frac{mx}{m} = x \qquad (A1.5)$$

$$\frac{y}{m} - \frac{c}{m} = x \qquad (A1.6)$$

2   Express y in terms of x, in:

$$3xy + 7y = x + 5 \qquad (A1.7)$$

Start by recognising that y is common to both terms on the LHS, and can therefore be factored out:

$$y(3x + 7) = x + 5 \qquad (A1.8)$$

The next stage is then, as before, to divide both sides by $(3x + 7)$ to give:

$$y = \frac{(x + 5)}{(3x + 7)} \tag{A1.9}$$

which is as far as this equation can be taken.

Algebra of complex numbers

Complex numbers are dealt with in Chapter 6. Here the only fact of relevance is that a complex number is of the form:

$$z = a + jb \tag{A1.10}$$

where $j = \sqrt{-1}$, that is $j^2 = -1$, and a and b can be numbers or algebraic symbols such as x or y. The algebra of complex numbers proceeds as before.

Examples

1 Add two complex numbers, $z_1 = a + jb$ and $z_2 = c + jd$, together:

$$z_1 + z_2 = (a + jb) + (c + jd) = (a + c) + j(b + d) \tag{A1.11}$$

The brackets around $a + c$ are not strictly necessary, but they are included here to show that not only can terms in j be collected together, but so also can the remainder, and the final expression then shows clearly that the result of the addition is another complex number.

2 Multiply the same two complex numbers together, that is:

$$z_1 z_2 = (a + jb)(c + jd) \tag{A1.12}$$

Each term in each bracket now has to be multiplied by each term in the other bracket,

$$z_1 z_2 = a \cdot c + jb \cdot c + a \cdot jd + j^2 b \cdot d$$

$$z_1 z_2 = (ac - bd) + j(ad + bc) \tag{A1.13}$$

where use is made of the fact $j^2 = -1$. The final result is another complex number.

3 Express y in terms of x:

$$(a + jb)y = mx + cy \tag{A1.14}$$

where a, b, m and c are constants. In principle this is no different from the other examples of algebraic manipulation. Start by collecting the terms in y, so subtract cy form both the LHS and RHS:

$$(a + jb)y - cy = mx$$
$$(a - c + jb)y = mx \tag{A1.15}$$

Then divide each side by $(a - c + jb)$:

$$y = \frac{mx}{(a - c + jb)} \tag{A1.16}$$

This is clearly a complex number, but it is not of the simple form described above. Is it possible to transform it anyway? Fortunately, yes. Make the substitution $d = a - c$ so that the complex number $d + jb$ appears on the bottom line. Now, if $d + jb$ is multiplied by $d - jb$ something peculiar happens:

$$(d + jb)(d - jb) = d^2 + jb \cdot d - jb \cdot d - j^2 b^2 = d^2 + b^2 \tag{A1.17}$$

The middle products cancel each other out leaving a real number. Thus if the RHS is multiplied by $(d - jb)/(d - jb)$, which is equal to unity, the total value of the RHS does not change so there is no need to do anything to the LHS, but the complex number on the bottom disappears. Multiplying one side of the equation by unity in this manner is equivalent, in the analogy of the balance, to rearranging the contents of one pan slightly without changing the amount of material. Balance is still maintained but the appearance is altered, that is:

$$\frac{mx}{(d + jb)} \times \frac{(d - jb)}{(d - jb)} = \frac{mx(d - jb)}{(d^2 + b^2)} \tag{A1.18}$$

Therefore,

$$y = \frac{dmx}{(d^2 + b^2)} - j\frac{bmx}{(d^2 + b^2)} \tag{A1.19}$$

which is recognisable again as a complex number. This process is called *rationalisation* of a complex number, and is a standard algebraic technique. Finally, $(a - c)$ can be substituted back into the equation for d, but it is not strictly necessary.

2 Logarithms

The logarithm of a number is defined as the power to which the base must be raised to give the number. Hence, for logarithms to base 10, we have:

$$\log_{10}(x) = y \tag{A2.1}$$

and

$$10^y = x \tag{A2.2}$$

Hence, $\log(10) = 1$, $\log(100) = 2$, $\log(1000) = 3$, and so on. For very small numbers $\ll 1$, the logarithm is negative. For example, $\log(0.1) = -1$, $\log(0.01) = -2$, and so on. Therefore, by using a logarithmic axis data extending over the range 10^{-2} to 10^3, for example, is effectively plotted on a scale extending from -2 to 3. For exponential functions, such as

$$e^y = x \tag{A2.3}$$

the *natural*, or *Naperian*, logarithm is used, written as:

$$\ln(x) = y \tag{A2.4}$$

The old fashioned way of taking logs involved looking them up in specially produced books of tables, accurate to four or five decimal places depending on your choice. Today, every calculator and software package capable of manipulating data includes logarithms, both to base 10 and Naperian, among its functions.

It follows from the law of additive indices, that is:

$$x^a x^b = x^{a+b} \tag{A2.5}$$

that the log of two numbers, say $c = x^a$ and $d = x^b$, multiplied together is:

$$\log(c \cdot d) = \log(c) + \log(d) \tag{A2.6}$$

and that the log of a number raised to some power is given by:

$$\log(x^n) = n \log(x) \tag{A2.7}$$

3 Differentiation of common functions

Suppose a general function of a power-law form,

$$y = ax^n \tag{A3.1}$$

is plotted as a smooth curve (Figure A3.1a). Suppose further that the curve can be approximated by straight line segments, as in Figure A3.1b. If each straight line segment extends over an interval Δx with a corresponding interval in the y direction of $\Delta y = y(x + \Delta x) - y(x)$, the gradient over this straight line segment is given by:

$$m = \frac{y(x + \Delta x) - y(x)}{(x + \Delta x) - x} = \frac{y(x + \Delta x) - y(x)}{\Delta x} \tag{A3.2}$$

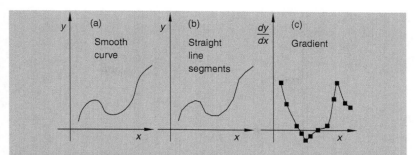

Figure A3.1 *Schematic representation of a curve as a series of straight-line segments from which a slope can be calculated*

Clearly, from equation (A3.1),

$$y(x + \Delta x) = a(x + \Delta x)^n = a\left[x\left(1 + \frac{\Delta x}{x}\right)\right]^n \tag{A3.3}$$

The term in $(1 + \Delta x/x)^n$ can be expanded using the binomial theorem, which gives the mathematical form for the expansion of a binomial (the sum of two numbers) to higher powers. In so far as this is a theorem (first published posthumously in 1665 by Pascal) and therefore a fact, the detailed mathematics of its derivation is of no concern to us as physicists. It is a tool; we take it and use it but we don't need to understand it. From the binomial expansion, then,

$$(1 + x)^n = 1 + nx + \frac{n(n-1)x^2}{2!} + \frac{n(n-1)(n-2)x^3}{3!} + \cdots + \frac{n!}{(n-r)!r!}x^r \tag{A3.4}$$

Only the first two terms are actually of interest. If $\Delta x/x \ll 1$, then $(\Delta x/x)^2$ is negligible and all higher powers of $\Delta x/x$ are also of no consequence (for example, $0.01^2 = 0.0001$, and $0.01^3 = 10^{-6}$). Hence:

$$y(x + \Delta x) \approx ax^n\left[1 + \frac{n\Delta x}{x}\right] \tag{A3.5}$$

and,

$$y(x + \Delta x) - y(x) \approx \left[ax^n + \frac{nax^n\Delta x}{x}\right] - ax^n \tag{A3.6}$$

The gradient becomes, from (A3.2):

$$m = \frac{\Delta y}{\Delta x} = nax^{n-1} \tag{A3.7}$$

In the usual notation, we write,

$$\frac{dy}{dx} = nax^{n-1} \tag{A3.8}$$

This is exact for a power-law dependence. If the function is a polynomial containing additive terms in different powers, each term is differentiated separately and the total differential is the sum of all the individual terms. For other functions, such as sine, cosine, exponential or logarithm, the differential takes a different form. The functions

$sin(x)$, $cos(x)$, and e^x-can be expressed as series in increasing powers of x and each term in the series can be differentiated in turn to show that:

$$\frac{d}{dx}[\sin(x)] = \cos(x) \tag{A3.9}$$

$$\frac{d}{dx}[\cos(x)] = -\sin(x) \tag{A3.10}$$

$$\frac{d}{dx}[e^x] = e^x \tag{A3.11}$$

and

$$\frac{d}{dx}[\ln(x)] = \frac{1}{x} \tag{A3.12}$$

Derivation of equation (A3.12) is quite long-winded, so very often the differential is given without explanation.

If the argument of a function is itself a function, for example ax, the chain rule can be used. For example,

$$\frac{d}{dx}[e^{ax}] = \frac{d}{d(ax)}[e^{ax}]\frac{d}{dx}[ax] = ae^{ax} \tag{A3.13}$$

There are also rules to give the differential of a product of functions or a quotient:

$$\frac{d}{dx}[u(x)v(x)] = v(x)\frac{d}{dx}[u(x)] + u(x)\frac{d}{dx}[v(x)] \tag{A3.14}$$

$$\frac{d}{dx}\left[\frac{u(x)}{v(x)}\right] = \frac{v(x)(d/dx)[u(x)] - u(x)(d/dx)[v(x)]}{v(x)^2} \tag{A3.15}$$

The differential of a function such as the tangent, $\tan(x) = \sin(x)/\cos(x)$, can be evaluated thus as:

$$\frac{d}{dx}\left[\frac{\sin(x)}{\cos(x)}\right] = \frac{\cos^2 x + \sin^2 x}{\cos^2 x} = \frac{1}{\cos^2 x} \tag{A3.16}$$

▶ Further reading

There are a number of books that deal with basic calculus, and in particular you may find the following helpful:

Basic Calculus: From Archimedes to Newton to its Role in Science, Alexander J. Hahn, Springer-Verlag, 1998.

Brief Calculus, W. A. Armstrong and D. Davis, Prentice Hall, 1999.

Mathematics, a Practical Odyssey, D. B. Johnson and T. A. Mowry, Brooks/Cole (USA), 2003, 5th edn.

Index